The Behavior of Gonadectomized Rhesus Monkeys

Contributions to Primatology

Vol. 20

Series Editor: *F.S. Szalay,* New York, N.Y.

Associate Editors: *P. Charles-Dominique,* Brunoy;
E. Delson, Bronx, N.Y.; *H. Kuhn,* Göttingen;
W.P. Luckett, Omaha, Nebr.; *J. Oates,* New York, N.Y.

Founding Editors: *A.H. Schultz* †, Zürich;
D. Starck, Frankfurt am Main

S. Karger · Basel · München · Paris · London · New York · Tokyo · Sydney

The Behavior of Gonadectomized Rhesus Monkeys

James Loy, Kent Loy, Geoffrey Keifer, Clinton Conaway

Department of Sociology and Anthropology, University of Rhode Island, Kingston, R.I.

46 figures and 17 tables, 1984

⊕ KARGER

S. Karger · Basel · München · Paris · London · New York · Tokyo · Sydney

Contributions to Primatology

National Library of Medicine, Cataloging in Publication
    The Behavior of gonadectomized rhesus monkeys
    James Loy ... [et al.]. – Basel; New York: Karger, 1984
    (Contributions to primatology; v. 20)
    1. Castration – veterinary   2. Macaca   3. Sex Behavior, Animal   I. Loy, James   II. Series
    W1 CO778UP v. 20 [QY 60.P7 B419]
    ISBN 3–8055–3795–6

Drug Dosage
    The authors and the publisher have exerted every effort to ensure that drug selection and dosage
    set forth in this text are in accord with current recommendations and practice at the time of
    publication. However, in view of ongoing research, changes in government regulations, and the
    constant flow of information relating to drug therapy and drug reactions, the reader is urged to
    check the package insert for each drug for any change in indications and dosage and for added
    warnings and precautions. This is particularly important when the recommended agent is a new
    and/or infrequently employed drug.

©   Copyright 1984 by S. Karger AG, P.O. Box, CH–4009 Basel (Switzerland)
    Printed in Switzerland by Thür AG Offsetdruck, Pratteln
    ISBN 3–8055–3795–6

Contents

Acknowledgments

The study reported here took ten years to complete. Like any lengthy study, it profited from the assistance, suggestions, and criticisms of a great many people. Most of the people who provided help over the years are included in the following list, but undoubtedly a few names have been inadvertently omitted. To everyone, named and unnamed, who contributed to the project, we extend our sincere thanks. Any remaining faults are ours alone.

Heading the list of individuals deserving thanks is Donald Patterson. Don was responsible for data collection between July 1974-May 1975, and by all rights should be a co-author of this study. Unfortunately, we lost track of Don's whereabouts in the last few years and were without the benefit of his help as we assembled the final report.

Among the numerous people who provided help during the data collection phase (1971-1976) were the staff members of the La Parguera primate colony (Caribbean Primate Research Center), especially Jacinto Rosado, Jose Flores, Israel Cordero, Abraham Rosado, and Americo Santiago. Donald S. Sade brought the monkeys to La Parguera from Cayo Santiago and made valuable suggestions about data collection procedures. C. Jean DeRousseau provided detailed genealogical information about the animals. Helpful members of the CPRC's Sabana Seca staff included William T. Kerber, who performed several of the gonadectomies and provided general veterinary support throughout the project, and also Carlos Torres, Oscar Lopez, and Irma Martinez.

John Vandenbergh performed several of the castrations during a visit to La Parguera. On another occasion, he assisted Jerry Hulka with follow-up surgery for one of the experimental monkeys.

Robert W. Goy and Jerry Robinson of the Wisconsin Regional Primate Research Center were helpful throughout the study. Dr. Robinson was in charge of all hormone assays for the project, and Dr. Goy was instrumental in providing JDL with the opportunity for a period of intensive data analysis at the WRPRC. Other Wisconsin Center staff members or students who deserve thanks for their assistance include William Bridson, Kim Wallen, Larry Jacobsen, Conrad Brown, Peter Goy, and Jan Thornton. Steve Suomi and Gordon Stephenson of the University of Wisconsin gave statistical advice. Special thanks go to Barbara and Bob Goy for their hospitality.

Many people at the University of Rhode Island helped to bring the project to a successful completion. The Coordinator of Research, N. McL. Sage, and his staff, and Betty Jones of the Department of Sociology and Anthropology, assisted with grant administration. Richard Pollnac, John Poggie, and Richard Gelles all provided guidance in the use of statistics. Several URI students, including Pauline Valcourt, Mary Lou Hurlbut, and Noel Brennen, helped with data analysis. Robert Farrell produced the beautiful drawings of the monkeys found on the cover and throughout the report. Very special thanks go to C. B. Peters for encouragement during the last years of the project and for help revising the report, and to Roseann Kane for her dedicated work typing several copies of the manuscript. Finally, thanks go to the URI administration for approving the sabbatical leave during which most of the report was written.

   The project was given general support throughout the data collection phase by its host, the Caribbean Primate Research Center. An affiliate of the Medical Sciences Campus of the University of Puerto Rico, the CPRC was funded by contract NIH-DRR-71-2003 from the Division of Research Resources, NIH. Two years of support specifically for the gonadectomy study was provided by grant number GB 44252 from the National Science Foundation. Grant number 1-R03-MH31982-01 from the National Institute of Mental Health provided support for two months of data analysis at the Wisconsin Regional Primate Center during the summer of 1979. Finally, publication costs for this report were paid by a grant from the University of Rhode Island Foundation, and contributions from the URI Sociology and Anthropology Department and the Coordinator of Research, URI.

I. Introduction

Over the past century, the close evolutionary relationships between humans, monkeys, and apes have been well documented. Due to their phylogenetic ties, all anthropoid primates show broad similarities in anatomy, physiology and behavior--a situation which has led to the widespread use of nonhuman primates as "models" in research into a variety of human problems. Within the biomedical sciences, for example, primates have been used as models in studies of the sexual differentiation of the brain [53, 62-64, 127, 136], the biology of reproduction [42, 66, 102], and numerous pathologies and diseases. Within the social sciences, primates have served as models in studies of aggression [79], sexual behavior [4, 56, 82, 87, 102, 111], socialization [31, 72], and a host of other areas. Finally, primate data have a long history of use in attempts to reconstruct the evolution of human society and behavior [39, 89, 106, 161, 178, 184].

One of the earliest and most influential studies of primates was ZUCKERMAN's <u>The Social Life of Monkeys and Apes</u> [184]. ZUCKERMAN summarized much of what was known at the time about primate behavior and social organization, and put forward the hypothesis that primate societies are ultimately based on uninterrupted sexual attraction between males and females. Other social factors, such as dominance rank among males, were recognized as important behavioral determinants, but sexual attraction was seen as "...providing the bonds that hold (males and females) together in permanent bisexual associations" (p. 147). Reflecting his training in the biological sciences, ZUCKERMAN traced primate sexual behavior to what he saw as its physiological bases, the gonadal hormones. He compared the mating behavior of monkeys and apes with that of non-primate mammals, and provided a hormonal explanation for his thesis that higher primate females mate throughout their menstrual cycles.

Over the years, ZUCKERMAN's ideas have prompted numerous field and laboratory studies, the results of which have not always supported the original hypotheses [90]. Nonetheless, his basic thesis concerning the importance of sexual attraction and interaction has not been disproved to date [98], and, as noted by ROWELL [144], "...his synthesis still remains a stimulating cause of argument" (p. 111).

The study described in this book was designed as yet another attempt to explore the area framed by ZUCKERMAN's hypotheses. In an effort to remove the key variable of sexual attraction from a rhesus monkey (<u>Macaca</u> <u>mulatta</u>) society--so that

we could observe the resulting behavior and social structure--
we performed pre-/peri-pubertal gonadectomies on all members of
a captive group. We expected that this manipulation would produce a "sexless" society, but, as shown in the following chapters, that expectation--like many others--proved to be somewhat
in error.

Our four and one-half year study produced a wealth of data
concerning rhesus behavior and social organization and the ways
these variables are influenced by the gonadal hormones. In addition, our project produced a considerable amount of information on the juvenile stage of social development and the behavioral transition that takes place between that stage and adulthood.

Because of its length, our project developed irregulari-
ties in design: four subjects died; one female continued menstrual and hormonal cycles after being ovariectomized; etc.
Although we probably would arrange certain things differently
if we had the project to do over, in many instances the irreg-
ularities contributed to interesting and important discoveries.
Thus, they proved to be valuable, if unintentional, additions
to the original design.

The chapters which follow fit into three broad categories.
First, in chapter two we discuss the history of our study and
our data collection procedures. Second, in chapters three
through nine we present data on the anatomical and behavioral
effects of gonadectomy. Finally, the body of data is followed
by two chapters in which we synthesize our findings and evaluate their significance.

II. History of the Study and Methods of Data Collection

Formation and Pre-operative History of Groups

Our study began on 18 November 1971, when 21 yearling
rhesus monkeys (Macaca mulatta) were transported by Dr. D.S.
Sade from the Caribbean Primate Research Center's Cayo Santiago
(Puerto Rico) colony to its sister colony at La Parguera. All
11 males and 10 females in this first shipment had been born
into the same Cayo Santiago social group (Group A) during the
1970 birth season (fig. 1), and their ages on arrival at La
Parguera ranged from about 17.5 to almost 22 months (table I).
These animals were released into a 0.2 ha corral on La Cueva
island and were provided with Wayne Monkey Diet and water ad
libitum. Observations of the yearlings' behavior began on 19
November.

On 3 December 1971, 12 additional animals (six males and
six females) were received from Cayo Santiago. Like the earlier arrivals, these monkeys had been removed from Cayo Santiago's Group A (fig. 1), but were somewhat older, having been
born during the 1969 birth season. Their ages on arrival
ranged from just over 30 to about 33.5 months (table I). These
two-year-old animals were released into the corral containing
the Group A yearlings.

Between November 1971 and late January 1972, a short behavioral study of the group of 33 juveniles was conducted
[100]. Ad libitum samples [5] of the animals' behaviors were
collected during morning hours, Monday through Friday. All
monkeys were tattooed, ear-notched and dye-marked for identi-
fication.

In preparation for the gonadectomy study, the 33 juveniles
were divided into a control group of 16 animals and an experimental group of 17 animals on 25 January 1972. The following
guidelines were used to determine group assignments (table I).

1. As nearly as possible, the groups were matched for the
ages and sexes of their members. Each group contained six
monkeys (three males and three females) born in 1969 and at
least 10 monkeys (five males and five females) born in 1970.
A sixth 1970-male was assigned to the experimental group,
bringing its total to 17 animals.

2. The groups were matched for the dominance ranks of
their members as determined during the November 1971-January
1972 study [100].

3. The 10 pairs of siblings were all split, with sibs go-
ing to opposite groups.

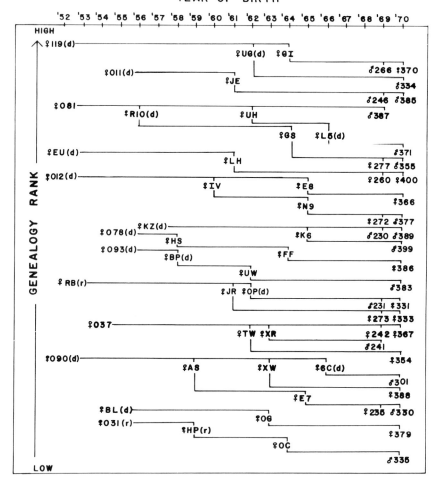

YEAR OF BIRTH

Fig. 1. Cayo Santiago Group A matrilines. This figure shows the genealogies present in 1968-69 and their relative dominance ranks. Only those monkeys born in 1969 and 1970 were used for the gonadectomy study. The symbols (d) and (r) indicate respectively animals that had died or had been removed from Group A prior to 1971. Figure from LOY and LOY [100] with permission of Alan R. Liss, Inc.

Table I. Composition of study groups*

Controls			Experimentals		

1969 birth crop

Name	Date of birth	Approximate rank**	Name	Date of birth	Approximate rank
Females			Females		
277	29 Mar.	5	260	6 Apr.	4
242	17 Feb.	10	272	20 May	6
235	1 Apr.	11	273	18 Feb.	9
Males			Males		
266	14 Apr.	1	246	13 Feb.	2
230	4 Mar.	7	241	4 Mar.	12
387	9 Apr.	3	231	21 Feb.	8
Mean rank = 6.2			Mean rank = 6.8		

1970 birth crop

Females			Females		
400	11 Apr.	5	367	6 Mar.	15
333	26 Jan.	14	370	10 Feb.	1
331	7 Mar.	12	366	19 May	7
379	9 Mar.	20	386	14 Feb.	11
388	29 May	19	354	22 Feb.	16
Males			Males		
377	25 May	8	355	4 Apr.	6
385	19 Mar.	3	330	29 Mar.	17
399	29 Mar.	10	389	21 Feb.	9
334	22 Feb.	2	335	25 Jan.	20
383	7 Feb.	20	371	4 Mar.	4
			301	2 Mar.	18
Mean rank = 11.3			Mean rank = 11.3		

*   Split sibling pairs: 277-355, 242-367, 235-330, 266-370, 230-389, 260-400, 272-377, 246-385, 231-331, 273-333.
** "Within age-class" dominance ranks determined prior to group formation [100].

4. When possible, animals who were seen to display adult sexual behavior (i.e. participate in a heterosexual mount series) and/or adult sexual cycling (as shown primarily by menstrual bleeding and secondarily by sex skin reddening and swelling) between November 1971-January 1972 [100] were assigned to the control group.

Following group assignment, the experimental monkeys were moved to a newly constructed 0.09 ha corral. The controls were retained in the original corral until 11 May 1972, when they were transferred to their own new 0.09 ha enclosure adjacent to that of the experimental monkeys. The two small corrals were nearly identical (fig. 2). Each was constructed of sheet-metal and chainlink fencing and allowed visual, auditory, olfactory, and limited tactile communication with the free-ranging rhesus monkeys of La Cueva island. Each corral was equipped with three sheet-metal shelters for shade and climbing, a set of metal "trees," a feeding station, and a water dispenser. The animals could hear loud vocalizations from the other group and, when standing bipedally atop the metal "trees," might have been able to see members of the other group who were also gazing bipedally from atop their "trees." Other visual communication between the two groups was prevented by solid metal walls separating the corrals. Some olfactory communication between the groups might have occurred, but we have no reason to think that it was an important factor during the study.

Between 26 January and 21 March 1972, a short period of pre-operative observation was conducted on the newly formed groups (table II). At the completion of these observations, all of the experimental monkeys were gonadectomized.

Gonadectomies

All of the experimental monkeys (with one exception to be discussed below) were either bilaterally castrated or bilaterally ovariectomized between 23 March and 7 April 1972 (table III). Standard sterile procedures were followed during the operations, and all animals recovered without incident after a few days of rest in individual cages. All castrations were pre-pubertal (between 2.0-3.1 years of age and prior to noticeable testicular descent). All ovariectomies were either pre- or peri-pubertal (between 1.8-3.1 years of age) (table III). Several of the experimental females had shown some sex skin swelling and coloration prior to surgery, and three females, numbers 354, 370 and 273, may have shown vaginal bleeding [185]. No surgical manipulations of any sort were performed on the control animals at this time.

The gonadectomies were complicated by female 370, who required two operations for bilateral removal of her ovaries. On 23 March, 370's right ovary was removed and we tried unsuccessfully to locate her left ovary. 370 was allowed to recover from this operation, put back in her group for a few weeks, and

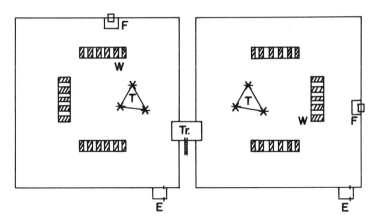

Fig. 2. Diagram of the corrals used during the study. Each corral was approximately 0.09 hectares in area. F feeder, W water, T metal trees, E entrance, Tr tower. Sheet-metal shelters are cross-hatched.

Table II. Study phases and observation hours

Phase	Ad libitum sampling		Scan sampling
Pre-operative: all 33 juveniles together (19 Nov. 1971- 21 Jan. 1972)	117.5 h*		---
Pre-operative: controls and experimentals separated (26 Jan. 1972- 21 March 1972)	Controls: Experimentals: Total:	72.2 h 73.7 h 145.9 h	--- --- ---
Post-operative (27 March 1972- 21 May 1976)	Controls: Experimentals: Total:	1041.3 h 1028.9 h 2070.2 h	254.25 h 253.50 h 507.75 h

Grand total ad libitum + scan sampling = 2841.35 h

* Two-year-olds only observed for 95.3 h [100].

Table III. Ages at gonadectomy

Sex	Date of operation	Age*
Males		
246	23 March 1972	3.1 years
241	23 March 1972	3.1
231	23 March 1972	3.1
371	7 April 1972	2.1
330	7 April 1972	2.0
301	7 April 1972	2.1
389	7 April 1972	2.1
335	7 April 1972	2.2
355	7 April 1972	2.0
Females		
260	23 March 1972	3.0
272	23 March 1972	2.8
273	23 March 1972	3.1
370	23 March 1972**	2.1
370	11 May 1972***	2.2
366	24 March 1972	1.8
386	24 March 1972	2.1
367	23 March 1972	2.0
354	24 March 1972	2.1

   * Ages rounded to nearest tenth of a year.
  ** Right ovary removed.
*** Left ovary removed.

on 11 May 1972, a second operation was performed during which her left ovary was found and removed. Although we thought that 370 was completely agonadal after her second laparotomy, she showed a recovery of some sex skin swelling and coloration during August 1972. On 11 September 1972, 370 was caught and laparotomized for the third time. This operation revealed no obviously ovarian tissue, but to be safe, the stub of her right fallopian tube was removed. Unfortunately, even her third operation failed to completely extinguish 370's sexual cycles. She showed three days of vaginal bleeding in late October 1972, and then continued to show regular menstrual and sex skin color cycles for the duration of the study.

370 was the highest-ranking experimental female. We felt that removing her from the study might do more harm (by disrupting the behavior and social relations within her group) than retaining her and sacrificing the completely agonadal make-up of the experimental group. 370 was retained, and during data analysis, we attempted to evaluate the effects of this cycling female on her gonadectomized groupmates.

Post-operative History of Groups

Observations of both groups were resumed immediately after the initial gonadectomies and continued without substantial interruption for the remainder of the study. The controls were allowed to breed normally during 1973 and 1974, and this resulted in the addition of several infants to that group. The intention of allowing births within the control group was to make that group as "normal" as possible, so that we could realistically measure the behavioral effects of gonadectomy. By early 1974, however, we realized that we were losing the ability to control for the infants' presence and began a program of infant removal. The four infants (two males and two females) born during 1973 were removed on 4 February 1974, and each of the five infants born in 1974 was removed as soon after birth as possible. In addition, control female 400's three to four month fetus was aborted in June 1974. On 30 May 1974, all control males were bilaterally vasectomized to prevent further pregnancies.

Other important post-operative events included the deaths of four monkeys and the apparent resumption of sexual cycling by a few ovariectomized females two years after their surgeries. Three of the monkeys who died were from the experimental group. Castrated male 389 died from an intestinal infection in October 1972; ovariectomized female 354 died in December 1973 due to a blocked digestive tract; and female 366 died due to unknown causes in July 1974. Control female 379 probably choked or suffocated after swallowing or inhaling dirt on 31 May 1973 as she was recovering from anesthesia after being handled.

Four experimental females showed signs of the resumption of sexual cycling in early 1974. Females 386 and 273 showed increases in their sex skin colors, while females 367 and 366 showed apparent vaginal bleeding. All four animals were given second operations between 31 May and 19 June 1974, and small amounts of "possible" ovarian tissue were removed from each female. These operations seemed to halt the recrudescence of sexual cycling, although 367 did show short periods of vaginal bleeding seven months (January 1975) and 19 months (January 1976) later.

Obviously, the history of the study is long and complicated. For the reader's convenience, the major manipulations and important events of the study are summarized in table IV.

Table IV. Histories of study groups

Date	Event
18 November 1971	21 yearling rhesus monkeys transferred from Cayo Santiago to La Parguera. Beginning of study.
3 December 1971	12 two-year-old monkeys transferred from Cayo Santiago to La Parguera.
25 January 1972	Formation of control and experimental groups. Experimentals moved to 0.09 ha corral.
23 March 1972	Bilateral castrations for males 246, 231, and 241. Bilateral ovariectomies for females 273, 272, 260, and 367. Right ovary of female 370 removed. (Operations by C.H. Conaway.)
24 March 1972	Bilateral ovariectomies for females 354, 386, and 366. (Operations by C.H. Conaway.)
7 April 1972	Bilateral castrations for males 389, 335, 330, 301, 371, and 355. (Operations by J.G. Vandenbergh.)
11 May 1972	Female 370's left ovary removed. (Operation by C.H. Conaway.) Controls moved to 0.09 ha corral adjacent to experimentals' enclosure.
4 July 1972	Beginning of scan sampling.
11 September 1972	Female 370 laparotomized to check for ovarian tissue, but none found; stub of right fallopian tube removed. (Operation by J. Hulka.)
19 October 1972	Castrated male 389 died.
October–December 1972	Control females 242, 235, 400 and 388 conceived.
April–June 1973	Four infants born into control group.
30 May 1973	Biannual bleeding and measuring program begun.
31 May 1973	Control female 379 died.
October 1973–February 1974	Control females 400, 331, 333, 242, 235, and 388 conceived.
8 December 1973	Experimental female 354 died.
4 February 1974	All 1973 infants removed from control group.
April–May 1974	1974 control infants removed from group immediately after birth.
30 May 1974	All control males bilaterally vasectomized. (Operations by W.T. Kerber.)
31 May 1974	Experimental female 367 laparotomized; "probable" ovarian nodules removed. (Operation by W.T. Kerber.)

Table IV. (continued)

Date	Event
19 June 1974	Experimental females 366, 273, and 386 all laparotomized. All produced "possible" ovarian tissue. Control female 400 laparotomized for removal of fetus. (Operations by W.T. Kerber.)
July 1974	Donald Patterson replaced Loys.
22 July 1974	Experimental female 366 died.
June 1975	Geoffrey Keifer replaced Patterson.
21 May 1976	Final day of observations.
25/26 May 1976	Final bleedings and measurements. End of study.

Data Collection Procedures: Behavior

Numerous individual and social behavior patterns were sampled throughout the study using a variety of techniques. Among the most important behavior patterns monitored were mounting and copulating, grooming, sitting-touching, fighting, play, and auto-eroticism. Each of these categories will be defined and discussed in detail in later chapters. For now, a general description of our data collection procedures will suffice.

As noted earlier, throughout the study behavioral data were collected during the morning hours, Monday-Friday. Occasionally there were deviations from this schedule due to inclement weather, interruptions, etc., but generally we worked five days per week and sampled the monkeys' morning activity peak. During the first year, data were collected by observers working inside the corrals at ground level. After 4 December 1972, observations were made from a tower constructed between the two corrals. Throughout the project, observation conditions within the corrals were excellent. Observers were always equipped with binoculars in case close inspection of an individual or an interaction was needed.

Four persons collected the behavioral data presented in this report: J. and K. Loy, November 1971-June 1974; D. Patterson, July 1974-May 1975; and G. Keifer, June 1975-May 1976. Periodic checks by the senior observer indicated a high degree of inter-observer agreement on behavioral descriptions and interpretations throughout the study.

Two primary data collection procedures were used during the project: ad libitum sampling and scan sampling. Ad libitum sampling [5] consisted of opportunistic data collection while monitoring an entire group. In other words, the observer made abbreviated notes describing interactions randomly observed across a social group. Although we attempted to sample the monkeys equally within the ad libitum framework, total observation times may have differed slightly across subjects. Typically, at least one hour of ad libitum sampling was done on each group every morning.

The ad libitum data were used for the construction of communication matrices, which produced valuable information about relationships between individuals. In addition, the ad libitum samples produced important descriptive information about certain infrequently observed behavior patterns, e.g. one male manipulating another's penis. The ad libitum data, however, did not provide a particularly reliable base for the accurate quantification of behavior. To overcome this deficiency, we added scan sampling to our data collection routine in July 1972.

Each scan sample (comparable to ALTMANN's [5] "all occurrences" samples) covered a 15 minute period during which the group under observation was closely watched for all instances of dyadic fighting, grooming, sitting-touching, mounting, and playing. As these behaviors were observed, they were scored on a checksheet according to the participants' sexes and their roles in the various interactions (e.g. winner or loser of a fight, groomer or recipient of a grooming bout). Interactions involving multiple (more than two) or unidentified monkeys were scored in the corresponding boxes. During scan sampling, neither the identities of the interacting monkeys nor a detailed description of the interaction were recorded. This allowed the observer to visually scan the group continuously and to record virtually all social and agonistic interactions. If an explanation of an interaction was necessary, the observer entered the appropriate information in the Remarks section of the checksheet after completion of the sample.

We attempted to scan sample each group twice per observation day between 0600-1200 hours. During most months, at least 25 scan samples (6.25 h) were collected on each group. The scan samples produced accurate estimates of the frequencies of important behaviors and the participation rates of the sexes in those behaviors.

These two procedures, ad libitum and scan sampling, produced all of the behavioral data contained in this report. A-cross the study, a grand total of 2333.6 h of ad libitum sampling was done. In addition, a total of 2031 scan samples were taken (507.75 h). Summing the two techniques produced a total observation time of 2841.35 h (table II).

Data Collection Procedures: Somatic and Physiological Measures

In order to understand more fully the behavioral effects of the gonadectomies, we periodically collected data on the sizes, weights, and hormonal states of the animals. Between

May 1973 and May 1976, all monkeys were caught for measurement and bleeding approximately every six months. Winter measurements (taken in late December or early January) coincided with the controls' mating season, and spring/summer measurements (taken in late May or early June) fell within the controls' non-mating season.

During each measurement session, all subjects were weighed, given a dental examination, and measured for body length and penis or clitoris width. In addition, all monkeys were bled and the blood samples were centrifuged to produce plasma. Plasma samples were shipped to the Wisconsin Regional Primate Research Center for anlaysis. Samples from females were assayed for estradiol and progesterone and those from males were assayed for testosterone.

The somatic measurements and hormone samples provided information about the effects of our surgeries on the growth patterns and physiological states of the experimental monkeys. Data from the controls verified that these animals were showing the annual hormonal cycles characteristic of rhesus monkeys.

Statistical Treatment of Data

Most of the behavioral data presented in this report were collected during scan sampling. The scan samples were summed monthly for each group, and the summed data were then used to calculate several behavioral indices (e.g. fights/female/h or mounts/male/h; definitions and methods of calculating these and other indices are presented in the various results chapters). Monthly values for the indices were used as the bases for inter-group comparisons, most of which were performed with one-way analyses of variances. Inter-group comparisons were done seasonally, across summed seasons, and/or across the entire post-operative period, as appropriate.

Fewer intra-group comparisons were made than comparisons between groups. Further, most intra-group tests were actually comparisons of male and female groupmates done with one-way ANOVAs. Very few intra-sex, intra-group tests were conducted. This was due to our feeling that despite the overall high level of inter-observer agreement, sufficient variation might have existed across the four observers in observation skills and behavioral interpretation to invalidate this sort of testing. For all statistical tests, alpha was set at 0.05.

We have only sketched our methods of data collection and analysis here. The chapters which follow will provide more detailed explanations and illustrations of the methods we employed. In the next chapter, we begin a presentation of the results of the study with an analysis of the somatic and hormonal data.

III. Somatic and Hormonal Measurements

Measurements and Methods

Twice each year, once during the controls' non-mating season and again during the mating season, all animals in both groups were captured for examination, measurement, and blood sampling. Capturing and handling was generally done early in the morning before the heat of midday, and was conducted entirely in the monkeys' home corral.

Each roundup began with the capture of the monkeys one by one. Most animals were netted by hand, but occasionally a few would try to hide in the wooden holding boxes and were caught by simply closing the crates' doors. After an animal was captured, it was anesthetized with approximately 0.5 cc of ketamine hydrochloride given IM. When the monkey was tranquil, it was weighed, measured, and had 8-10 cc of blood drawn from the saphenous vein or a femoral vessel with a heparinized syringe (see [28] for information on the lack of ketamine HCl effect on blood hormone levels). The monkey was then usually placed back into a holding crate and allowed to recover from the anesthetic.

The task of catching and processing all of the members of a group generally took two hours or more. This meant that the last monkeys measured and bled had either been held for an hour or so in a crate, or had watched from across the corral for that length of time as their groupmates were handled. In all, the capture and handling experience was probably fairly stressful for the monkeys, and while the stress probably did not significantly affect somatic measurements, it might have influenced hormonal values [36, 104, 132]. (There is, however, contradictory evidence on the effect of stress on testosterone levels [141].)

The following somatic measurements were made on all animals.

1. Weight: The tranquilized monkey was placed in a box or other container and then atop the scale. The weight of the container was later subtracted to produce the animal's weight.

2. Crown-rump length: The measurement was taken with the subject reclined along a rigid measuring board.

3. Width of glans penis (males): This measurement was taken with vernier calipers at the widest point of the glans and at 90° to the external orifice of the urethra.

4. Width of the clitoris (females): When the clitoris could be recognized, calipers were used to measure its maximum width. The measurement was taken perpendicular to the mid-line of the body.

Although each monkey's teeth were briefly examined during every roundup, dental measurements were not taken regularly. Toward the end of the study, however, we began to suspect that the two male populations differed significantly in canine lengths. Therefore, during the May 1976 roundup, we measured the lengths of the upper and lower right canines for all males. Measurements were made from the gum line to the tip of the tooth. All males were six to seven years of age when these data were collected, and canine growth should have been complete [81].

After all monkeys scheduled to be handled on a particular day had been processed, the blood samples (which had been kept cool since drawing) were centrifuged to produce plasma. Plasma samples were frozen prior to shipment to the Wisconsin Regional Primate Research Center. Under the direction of Dr. J.A. Robinson, the samples from females were assayed for estradiol and progesterone, and those from males were assayed for testosterone. The assay methods used at the Wisconsin Center have been described in various publications [20, 138].

Due to a variety of factors including somewhat imprecise measuring procedures, small sample sizes, and the possible influence of stress on hormonal values, only a few statistical tests were run on the somatic and hormonal data. In the following sections, we will primarily use gross inter-group comparisons to demonstrate the impact of gonadectomy on the anatomical development and hormone production of the experimental monkeys.

Results: Somatic Measurements and Growth

Males
The somatic data were grouped by six-month age blocks for analysis, with block 1 representing 36-41 months of age. The controls' data were averaged across all males, while data from the castrated males were divided according to year of birth (i.e. means for castrated males born in 1969 were calculated separately from those for 1970 males). The experimental males' data were divided in this fashion in order to check the two age-classes for different growth patterns resulting from different ages at castration (3.1 years for the 1969 males versus 2.0-2.2 years for the 1970 males) (table III). The older experimental males had probably begun their juvenile growth spurt a few months prior to castration [172].

With regard to body weight, the control males were consistently heavier than the experimental males between 42-89 months of age (fig. 3). Inter-group differences (comparing all controls with all experimentals) were generally about 1.4 kg. The question, however, as to whether or not these weight dif-

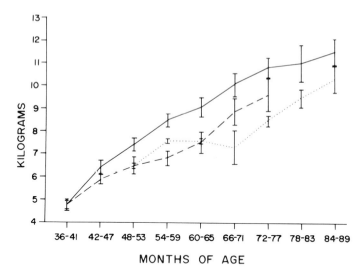

Fig. 3. Males' body weights. Mean weight ± SE. Control males (all) _____, castrated males born in 1969 ...., castrated males born in 1970 - - - - - - - -.

ferences were biologically significant remains open. The body weights shown by the control males and by both age-classes of castrated males (the latter subgroups did not appear to differ appreciably in weight gain patterns) were very close to the weights reported for intact, laboratory-reared rhesus males [172]. Overall, it appeared that castration depressed weight gain in males, but only slightly.

The results of the analysis of male crown-rump length were similar to those for body weight. The controls showed greater crown-rump lengths than either age-class of castrated males during eight of nine six-month blocks (fig. 4). Average inter-group difference was about 1.8 cm. Despite these differences, however, both age-classes of castrated males and the control males showed crown-rump lengths that agreed closely with data from intact, laboratory-reared males [172]. Interestingly, in contrast to body weight, the data on crown-rump length among the experimental males suggested an effect of age at castra-tion. The 1969 castrates exceeded the 1970 castrates by an average of just over 2 cm between 48-77 months of age. Over-all, castration appeared to slightly depress growth as measured by crown-rump length, with age at gonadectomy an important factor.

As expected, there were inter-group differences in penis width across all measurement periods (fig. 5). The control

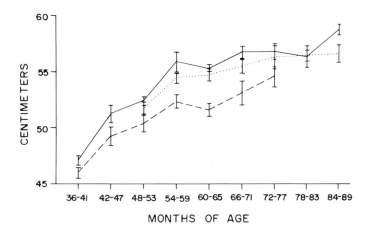

Fig. 4. Males' crown-rump lengths. Mean ± SE. Control males (all) _____, castrated males born in 1969 ...., castrated males born in 1970 - - - - - - - -.

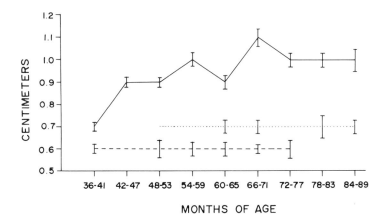

Fig. 5. Penis widths. Mean ± SE (missing SEs were less than 0.01). Control males (all) _____, 1969 castrates ...., 1970 castrates - - - - - - -.

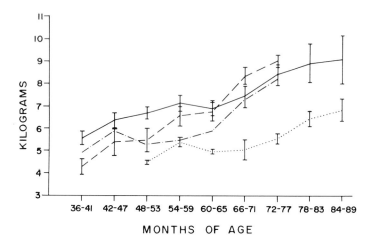

Fig. 6. Females' body weights. Mean ± SE. Control females (all) _____, experimental females born in 1969
.........., experimental females born in 1970 (excluding 370)
- - - - - - -, female 370 ---.---.---.---.---.

males' penises grew to an average width of 0.9-1.1 cm, while
those of the experimental males averaged 0.6-0.7 cm. In addition, age at castration may have had a slight effect on penis
width among the experimental males.

Finally, an analysis of the canine length data taken during the May 1976 roundup indicated that at six to seven years
of age the control males had significantly longer upper canines
(1.51 ± 0.09 [SE] cm versus 1.10 ± 0.05 cm; t test, p < 0.001)
and lower canines (0.92 ± 0.05 cm versus 0.74 ± 0.03 cm; t
test, p < 0.02) than the castrated males. Age at castration did
not have a discernable effect on adult canine lengths.

Females
The body weight data collected from the control females
are confounded somewhat by the inclusion of a few pregnant
animals in the first four six-month blocks (fig. 6). Block 1
(36-41 months of age) contains two females (388 and 400) who
were very near term. The data from these females were used
along with those from non-pregnant animals in the calculation
of mean female weight, but only after the pregnant females'
raw weights had been reduced by 13% (i.e. reduced by the average weight of the rhesus newborn [74]). Age blocks 2, 3, and
4 contain three, one, and two pregnant control females, respectively. The raw weights of these females, all of whom
were one to four months pregnant, were not adjusted, and therefore the control means for those age blocks are slightly in-
flated. Nevertheless, in spite of the complicating effects of

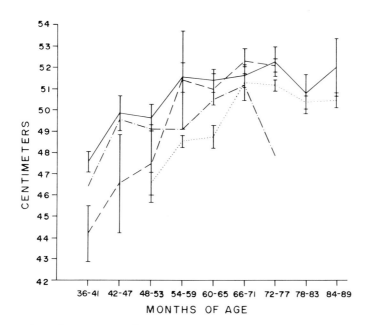

Fig. 7. Females' crown-rump lengths. Mean ± SE. Control females (all) _____, 1969 experimentals ...., 1970 experimentals (excluding 370) - - - - - - -, female 370 ---. ---.---.---. The measurement for female 370 at 72-77 months of age is almost certainly in error.

pregnant animals in the sample, the mean body weights of the control females agreed closely with data for laboratory-reared monkeys [172].

Also closely comparable to the lab data were the body weights of the younger (1970) experimental females (fig. 6). The weights of these females were all within one standard deviation of the lab values (with the exception of the mean from 72-77 months of age, which was over one SD above the lab means). In contrast, the mean weights for the three 1969 experimental females averaged 2.3 kg lower than like-aged controls and were consistently more than one SD below the lab means (with the exception of the 78-83 month block). Finally, the cycling experimental female (370) showed a pattern of weight gain very similar to that of the intact females and the 1970 experimental females.

With regard to crown-rump length, all categories of females (controls, 1969 experimentals, 1970 experimentals, female 370) closely approached the means for lab-reared animals (fig. 7) [172]. Once again, however, the 1969 experimental females consistently scored slightly lower than the other subjects of the present study.

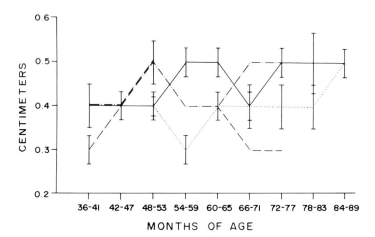

Fig. 8. Clitoris widths. Mean ± SE (missing SEs were less than 0.01 or represent an N of one female). Control females (all) _____, 1969 experimentals ...., 1970 experimentals (excluding 370) - - - - - - -, female 370 ---.---.---.---.---.---.---.---.---. Gaps in the data are due to the fact that on occasion it was impossible to measure accurately certain animals.

Finally, ovariectomy may have had a slight effect on clitoris width, but the data were inconclusive (fig. 8). All categories of females showed mean clitoris widths of 0.3-0.5 cm throughout the study.

Results: Plasma Hormone Levels

Males
The initial blood samples were taken in May 1973, during the controls' non-mating season (fig. 9). At that time, the 1969 control males were just over four years old and sexually mature [38, 134]. The 1970 control males had just passed three years of age, had recently completed or were in the process of completing testicular descent, and presumably were approaching adult levels of testosterone production [134, 140].

As shown in table V, both subgroups of control males experienced a general increase in testosterone production as they grew older. In addition, the control males showed the seasonal changes in testosterone production that are characteristic of rhesus monkeys in many environments (fig. 9) [59, 131, 150]. While the controls' mean testosterone values differed somewhat from the levels reported for certain lab or corral-housed populations [59, 131, 140] (but see WICKINGS and NIESCHLAG [177]

Table V. Plasma testosterone values (ng/ml)

Bleeding date	Age-class	Controls			Castrates		
		N	Mean	SE	N	Mean	SE
May 1973	'69 males	3	0.52	0.08			
	'70 males	5	0.29	0.05			
	all males	8	0.38	0.06	8	0.20	0.05
December 1973	'69 males	3	2.77	0.44			
	'70 males	5	2.86	0.66			
	all males	8	2.82	0.45	8	0.13	0.03
May 1974	'69 males	3	0.86	0.39			
	'70 males	5	0.76	0.14			
	all males	8	0.80	0.17	8	0.12	0.03
January 1975	'69 males	3	2.77	0.54			
	'70 males	5	3.03	1.03			
	all males	8	2.93	0.67	8	0.15	0.03
June 1975	'69 males	3	2.23	0.94			
	'70 males	5	1.61	0.30			
	all males	8	1.84	0.41	8	0.18	0.02
January 1976	'69 males	3	3.27	0.21			
	'70 males	5	3.52	0.52			
	all males	8	3.43	0.34	8	0.15	0.03
May 1976	'69 males	3	1.34	0.14			
	'70 males	5	1.78	0.62			
	all males	8	1.62	0.40	8	0.17	0.02

for lower laboratory values), the controls' non-mating season lows and mating season peaks were very similar in magnitude to those shown by free-ranging rhesus males on Cayo Santiago [150]. Given the environmental similarity between Cayo Santiago and La Parguera, it appears that the controls' testosterone production was quite normal for the conditions.

As expected, castration reduced the experimental males' plasma testosterone values to very low levels that probably reflected adrenal androgen production (table V; fig. 9) [137]. The castrated males showed neither an increase in testosterone levels with age, nor seasonal changes in their testosterone values. In addition, the castrated males failed to develop the hormone-dependent sex skin colors characteristic of intact males [164].

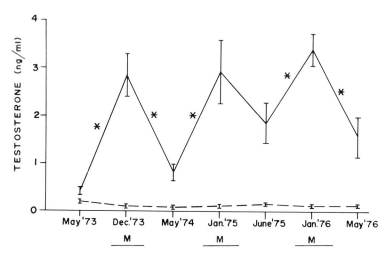

Fig. 9. Plasma testosterone values (ng/ml). Mean ± SE. Control males (all) _____, castrated males (all) - - - - - - - - -. M indicates the mating season in the control group. A significant difference between successive mean testosterone values for the controls is shown by an asterisk (*). Differences were evaluated with the Wilcoxon matched-pairs signed-ranks test or the sign test.

Females
    The hormonal data for the control females (and experimental female 370) are difficult to interpret without accurate information about the location of each female within her menstrual cycle on each bleeding date. In most cases, such information was not obtained. Therefore, for the control females the total range of values for each hormone will be given, as well as more specific information about the few values that could be assigned to a particular stage of the menstrual cycle (menstrual cycle divided as follows: follicular phase, menstruation plus first third of cycle; mid-cycle, middle third of cycle; luteal phase, final third of cycle).
    The controls' estradiol values showed an overall range of 17-244 pg/ml of plasma. Five values could be reliably assigned to the first third of the menstrual cycle, and they ranged from 35-75.4 pg/ml. Four mid-cycle values ranged from 25.4-128.3 pg/ml. All of these data are in good agreement with other reports of estradiol levels in intact rhesus females [124].
    Plasma progesterone values for the control females showed an overall range of 0.08-5.49 ng/ml. Five follicular phase values ranged from 0.30-1.83 ng/ml, while four mid-cycle values

Table VI. Hormone values for the experimental females

Bleeding date	All females except 370			Female 370*	
	N	Mean	SE		
A. Plasma estradiol (pg/ml)					
May 1973	6	12.2	0.52	16.5	(L)
Dec. 1973	6	13.3	2.35	226.5	(late F)
May 1974	5	9.5	0.88	72.5	(M)
Jan. 1975	5	10.2	1.17	63.0	(M)
June 1975	5	13.5	2.68	26.1	(U)
Jan. 1976	5	23.2	1.22	118.5	(M)
May 1976	5	20.8	0.64	123.0	(U)
B. Plasma progesterone (ng/ml)					
May 1973	7	0.68	0.15	0.43	(L)
Dec. 1973	6	0.49	0.10	0.56	(late F)
May 1974	6	0.36	0.07	1.19	(M)
Jan. 1975	5	0.36	0.06	5.03	(M)
June 1975	5	0.31	0.05	0.49	(U)
Jan. 1976	5	0.19	0.01	0.98	(M)
May 1976	5	0.17	0.03	2.94	(U)

* Letter following hormone value indicates approximate cycle
  stage: F = follicular, M = mid-cycle, L = luteal, U = stage
  uncertain.

ranged from 0.17-5.49 ng/ml. These results agree with other
studies of blood progesterone in intact rhesus females [77,
123, 124].

In contrast to the control females, the ovariectomized
animals showed consistently low values for both estradiol and
progesterone (table VI). The hormone levels found in our experimental females were in agreement with the results of other
studies of gonadectomized animals [77, 135], and presumably
reflected hormone secretion by the adrenal glands [135]. Female 370, who continued regular menstrual cycling after bilateral ovariectomy (N = 37 cycles; $\bar{X}$ cycle length = 38.7 $\pm$ 5.45
days, calculated following KAPLAN and MEIER [84]; median and
modal lengths of 27 and 24 days, respectively) [27, 74, 179],
showed hormonal ranges characteristic of intact females (table
VI). As a consequence of her normal hormonal cycles, 370
showed monthly fluctuations in sex skin color similar to those
of the control females [24, 40, 41, 185].

Discussion

The differences between the control and castrated males in body weight and length, penis width, and canine length were probably all related to the influence of testicular androgens on the intact males' growth, and the lack of significant androgen stimulation among the castrated males. Testosterone has been shown to augment growth in laboratory macaques [170, 173] and is thought to be at least partially responsible for the adolescent growth spurt in boys [22]. In addition, the pubertal increase in testosterone is correlated closely with increasing penis size in boys [55].

The effect of age at castration on the experimental males' crown-rump lengths could also have been related to testosterone stimulation. Rhesus males begin to experience a rise in plasma testosterone (from the generally non-detectable levels of immaturity) late in their third year or early in their fourth year of life [134]. The experimental males born in 1969 may have experienced sufficient testosterone stimulation prior to their castrations to have produced greater mean body length than that shown by the males castrated at a younger age.

The somatic data from the females are difficult to explain in terms of the hormonal effects of ovariectomy. The control females (and apparently experimental female 370 as well) were stimulated by high gonadal estrogens and presumably by gonadal testosterone [77]. Additionally, all females presumably received stimulation from adrenal testosterone [135]. Androgens are thought to stimulate growth in girls [22] and female monkeys [171], while estrogens are believed to inhibit growth [22]. Therefore, considering only gonadal secretions, the control females should have experienced both stronger growth stimulants and stronger growth inhibiting factors than their experimental counterparts. It appears possible that the combination of hormonal stimulation and inhibition experienced by the controls was essentially the same as that experienced (at lower absolute hormone levels) by the ovariectomized females, with the result that the growth patterns shown by the two female populations were very similar. Put differently, ovariectomy may not have altered significantly the ratio of androgen stimulation and estrogen inhibition, and therefore had only a small effect on growth. Left unexplained by this hypothesis are the low body weights shown by the 1969 experimental females, who were gonadectomized peri-pubertally.

Summary

Castration between 2.0-3.1 years of age had small but clear effects on several somatic measurements. Compared to intact controls, the castrated males were lighter, shorter from crown-to-rump, and had smaller glans penis widths and shorter adult canines. All of these differences were probably related to the stimulating effect of testosterone on growth and the loss of such stimulation among the castrated males. Perhaps because of early increases in testosterone, peri-pubertal castration had less effect on certain somatic measurements than did pre-pubertal castration.

Ovariectomy apparently had fewer somatic effects than castration. The experimental females were about as heavy and as long, and had about the same size clitorises as their intact counterparts. An exception to this generalization was the low mean body weight shown throughout the study by the 1969 experimental females. It appeared possible that peri-pubertal ovariectomy had more of a depressing effect on body weight than pre-pubertal removal of the ovaries.

The hormone levels shown by the control males and females (and the males' seasonal cycles in testosterone production) were normal for the species. In contrast, within the experimental group the males showed very low levels of testosterone and most females showed low levels of estradiol and progesterone, thus verifying the success of our gonadectomies. The one exceptional experimental animal was female 370, whose high hormone values verified the normalcy of her menstrual and sex skin color cycles.

With these data on growth and physiology as a background, we will now turn to the behavioral results of the study.

IV. Agonism and Dominance

Definitions and Methods

Fighting occurred frequently among both the control and
experimental monkeys. Following SADE [148], a fight was defined as an attack of any intensity followed by a flight of any
intensity. SADE also provided a catalog of rhesus aggressive
and submissive gestures which we modified slightly by deleting
two flight patterns (glance away and present) not considered
to be clear indicators of defeat [100].

During scan sampling, dyadic fights were recorded according to the combatants' sexes and roles in the interaction (i.e.
winner-loser). Fights involving more than two monkeys were
classified as "multiple" fights and scored separately, as were
fights involving unidentified participants. Fights involving
control group infants were noted during scan sampling, but were
not used in the following analysis.

At the end of each month, the scan sample data for each
group were summarized, and monthly means were calculated for
several behavioral indices. The agonistic indices were calculated as follows:

1. Fights/dyad/h: A group's total number of dyadic fights
between identified (and therefore sexed) participants was divided by the total number of dyads in the group.* The resulting number was then further divided by the month's total scan
sampling hours. This index provided an overall measure of
agonism within the group.

2. Male-male fights/dyad/h: The total number of fights
between males was divided by the number of male-male dyads in
the group. The resulting figure was then divided by the total
scan hours for the month.

3. Male-female fights/dyad/h: The total number of fights
in which a male defeated a female was first divided by the
number of heterosexual dyads in the group and then by the total
scan hours for the month.

4. Fights/male/h: Male-male and male-female fights were
summed. The resulting figure was then divided first by the
number of males in the group and then by the month's total scan
hours. This index provided a general measure of male aggressiveness, but was not an accurate measure of total male participation in agonism since it did not utilize data from heterosexual fights won by females.

* Total dyads calculated as follows: $\frac{N \times N-1}{2}$.

5. Female-female fights/dyad/h: This index was calculated similarly to index 2, only for females.

6. Female-male fights/dyad/h: Following the same format as index 3, this index quantified heterosexual fights won by females.

7. Fights/female/h: This index was calculated similarly to index 4, only for females.

8. Heterosexual fights/dyad/h: Numbers of male-female and female-male fights were summed. This total was first divided by the number of heterosexual dyads, and then by the scan hours for the month.

9. Isosexual fights/dyad/h: Male-male and female-female fights were summed. The total was first divided by the number of isosexual dyads in the group and then by total scan hours.

Monthly means for the nine indices were used as the basic data for inter-group statistical comparisons using one-way ANOVAs. Comparisons were made (1) across the entire post-operative period of scan sampling (July 1972-May 1976), (2) for each mating and non-mating season, and (3) for summed mating and non-mating seasons. Monthly means were also used to calculate grand means for seasons, summed seasons, and the total post-operative period.

Mating and non-mating seasons were defined by the behavior of the control monkeys, with mating seasons marked by observation or evidence of heterosexual copulation (see chapter V). The dates of the controls' reproductive seasons were as follows:

Mating 1 - August 1972 through January 1973
Non-mating 1 - February through September 1973
Mating 2 - October 1973 through March 1974
Non-mating 2 - April through July 1974
Mating 3 - August 1974 through March 1975
Non-mating 3 - April through September 1975
Mating 4 - October 1975 through April 1976

Both July 1972 and May 1976 were non-mating months and, while they were not included in any of the above seasons, data from these months were used for inter-group comparisons across summed non-mating seasons and across the total scan sample period.

Ad libitum observations of fights were made in addition to scan sampling. Ad libitum fights were described in prose notes and winners and losers were identified (fights resulting in a "draw" or involving unidentified monkeys were not analyzed). Data on fights won and lost were used to arrange the monkeys with regard to dominance. For example, if an animal head-bobbed toward a groupmate who cowered, the first monkey was declared the winner of the fight and (at least temporarily) the dominant member of the dyad. During analysis, the members of each group were assigned monthly dominance ranks based on their total number of subordinates (monkeys they were observed to defeat at least once, plus monkeys who showed submissive behavior without being threatened; see LOY and LOY [100] for a discussion of "unprovoked submissive behavior"). Animals with numerous subordinates were assigned high dominance ranks (low

numerical values), while those with few subordinates were rel-
egated to low ranks (high numerical values). Monthly dominance
rankings were tested for relationships with sex, size, aggressiveness, and (for males) testosterone level using Pearson
product-moment correlations.

Finally, we tracked dyadic relationships within each group
through the use of monthly dominance matrices [100, 148].
Changes in relationships (either partial changes or complete
reversals) were found by comparing matrices from successive
months, and were used to calculate a monthly Dominance Change
Score (DCS). Inter-group tests of dominance stability were
performed by analysis of variance.

Results: Pre-operative Agonism

Ad libitum data collected between November 1971 and 21
January 1972 indicated that the males who would later be se-
lected for the control group did not differ significantly from
pre-castrate males in mean fights/male/h (t test, p > 0.05).
Similarly, pre-control and pre-experimental females did not
differ significantly in mean fights/female/h (t test, p > 0.20;
all above data taken from LOY and LOY [100]).

Between 26 January and 21 March 1972, the control and experimental groups were separated, but all animals were still
intact. During this last pre-operative period, ad libitum data
indicated identical fighting frequencies in the two groups
(mean fights/monkey/h = 0.82). Inter-group differences in vic-
tories/h were non-significant for both males and females (Mann-
Whitney U-tests, p > 0.10 and p > 0.22, respectively).

Together, these two sets of pre-operative data suggested
that the control and experimental monkeys began the study with
equal potential for agonistic interactions.

Results: Post-operative Agonism (Scan Data)

Fights/dyad/h
Across the entire scan sampling period (July 1972-May
1976), the controls showed an average of 0.14 ± 0.01 (SE)
fights/dyad/h; a rate not significantly different from the experimentals' mean of 0.16 ± 0.01. When individual seasonal
means were tested, it was found that the groups differed significantly only during mating season 4 (October 1975-April
1976), when the experimentals outfought the controls (0.13 ±
0.01 vs. 0.10 ± 0.01; F = 5.13, df 1/12, p < 0.05). Inter-
group comparisons across the summed mating seasons and across
the summed non-mating seasons revealed non-significant differences.

These results showed that the controls and experimentals
were similar in overall frequencies of fighting. Other indices
indicated different contributions by males and females to each
group's agonistic behaviors.

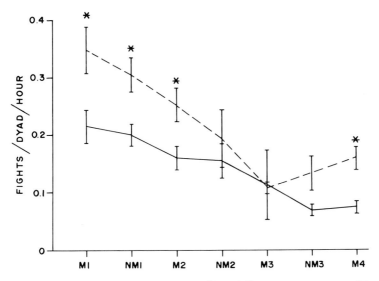

Fig. 10. Male-male fights/dyad/hour. Mean ± SE. Control males (all) _____, castrated males (all) - - - - - - -. M indicates the mating season; NM indicates the non-mating season. An asterisk (*) indicates a significant inter-group difference for that season.

Male-male fights/dyad/h
    The castrated males fought with each other significantly more often across the total post-operative period (i.e. the total scan sampling period) than did the control males (0.22 ± 0.02 vs. 0.14 ± 0.01 male-male fights/dyad/h; F = 14.06, df 1/90, p < 0.001). A seasonal comparison indicated significantly more male-male fighting among the castrates during four of the seven post-operative reproductive seasons (fig. 10), across the summed mating seasons (F = 6.29, df 1/50, p < 0.05), and across the summed non-mating seasons (F = 7.94, df 1/38, p < 0.01). We were not surprised to find that the castrated monkeys showed little seasonal change in male-male fighting (summed mating season mean of 0.21 ± 0.02 fights/dyad/h vs. summed non-mating season mean of 0.23 ± 0.02), but we had not expected to find a lack of seasonal change among the control males as well (summed mating season mean of 0.14 ± 0.02 male-male fights/dyad/h vs. summed non-mating season mean of 0.15 ± 0.02; see fig. 10).

Male-female fights/dyad/h
    The castrated males not only fought with each other more often than did the control males, but in addition, they defeated their ovariectomized groupmates significantly more frequent-

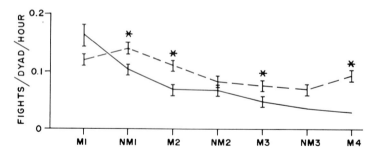

Fig. 11. Male-female fights/dyad/hour. Mean + SE (missing SEs were less than 0.01). Controls (all) _____, experimentals (all) - - - - - - -. M mating season, NM non-mating season. An asterisk (*) indicates a significant inter-group difference for that season.

ly than the control males defeated the intact females (total post-operative period: experimentals 0.10 ± 0.01 male-female fights/dyad/h vs. controls 0.08 ± 0.01; F = 8.88, df 1/90, p < 0.01). This inter-group difference did not hold across the summed mating seasons, even though the experimentals showed significantly higher values during three of the four individual mating seasons (fig. 11). In contrast, the gonadectomized monkeys showed a significantly higher value across the summed non-mating seasons (F = 4.90, df 1/38, p < 0.05), although they significantly exceeded the controls during only one individual season. Finally, both male populations failed to show seasonal changes in their frequencies of male-female fights/dyad/h (fig. 11).

Fights/male/h
The data presented above indicate that the castrated males were more aggressive than their control counterparts. This surprising finding was confirmed by a measure of fights/male/h and also by an independent measure of male aggressiveness taken from the ad libitum data.

Fights/male/h measured the overall victory rate for males regardless of opponents' sexes. This index provided a fairly precise measure of male aggressiveness since, in our experience, rhesus monkeys rarely threaten or attack an opponent they are unlikely to defeat. As noted earlier, fights/male/h does not provide a good measure of total male involvement in agonistic behavior because it does not utilize data from heterosexual fights won by females. Across the entire scan sampling period, the castrated males showed significantly more fights/male/h than the control males (1.49 ± 0.10 vs. 1.02 ± 0.10; F = 11.63, df 1/90, p < 0.01). This difference held true

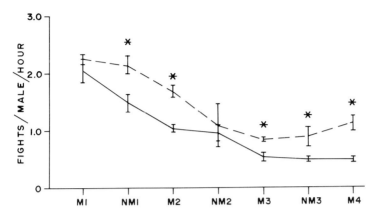

Fig. 12. Fights/male/hour. Mean ± SE. Control males
(all) _____, castrated males (all) – – – – – – –, significant inter-group difference shown by *.

during five of the seven post-operative reproductive seasons
(fig. 12), across the summed mating seasons (F = 6.33, df 1/50,
p < 0.05), and across the summed non-mating seasons (F = 5.10,
df 1/38, p < 0.05).
  The second measure of male aggressiveness was taken from
ad libitum data. Each month every animal's rate of victories/
subordinate/h was calculated. Any monkey that had been defeated by the subject in question at least once that month (or who
showed "unprovoked submissive behavior" toward the subject) was
counted as a subordinate. The number of hours used for calculations was the month's ad libitum sampling total for the
group. Across the entire post-operative period, the castrated
males averaged 0.11 ± 0.004 victories/subordinate/h, as compared to the control males' mean of 0.09 ± 0.003. The difference was significant (F = 19.69, df 1/92, p < 0.001). As shown
in figure 13, the significant overall difference in male aggressiveness was probably related to a steady post-operative
decline in aggressiveness among the control males that was not
matched among the castrated males.

  Female-female fights/dyad/h
  Over the entire post-operative period, the control females
fought among themselves significantly more frequently than the
ovariectomized females (controls 0.25 ± 0.01 vs. experimentals
0.19 ± 0.01 female-female fights/dyad/h; F = 12.10, df 1/90,
p < 0.001). The controls did not consistently show more inter-
female fighting than the experimentals, however. A seasonal
analysis revealed significant differences only during mating
seasons 1 and 4 (fig. 14). As a consequence, the two female
populations differed significantly across the summed mating

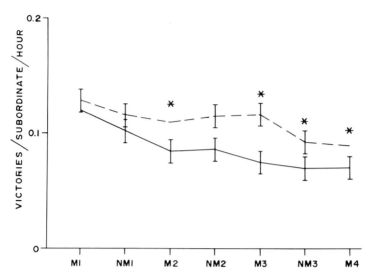

Fig. 13. Males' victories/subordinate/hour. Mean ± SE
(missing SEs were less than 0.01). Control males (all)
_____, castrated males (all) - - - - - - -, significant
inter-group difference shown by *.

seasons (controls 0.24 ± 0.01 vs. experimentals 0.17 ± 0.01;
F = 16.06, df 1/50, p < 0.001), but not across the summed non-
mating seasons.

Female-male fights/dyad/h
    Within both groups there were several instances of females
dominating males. As a consequence, heterosexual fights won by
females were not rare, although in both groups they occurred
less frequently than male wins over females (female victory
rate was 44% of the male victory rate among the controls and
24% among the experimentals). Across the entire post-operative
period, the control females showed significantly more female-
male fights/dyad/h than the experimental females (0.03 ± 0.003
vs. 0.02 ± 0.003; F = 4.44, df 1/90, p < 0.05). The inter-
group difference was neither large nor consistent, however.
The controls exceeded the experimentals during only one season
(mating 2) and there were non-significant differences across
the summed mating and across the summed non-mating seasons.

Fights/female/h
    Contrary to the relationship between the male populations,
the scan sample data suggested that the control females were
more aggressive than the experimental females. This was par-

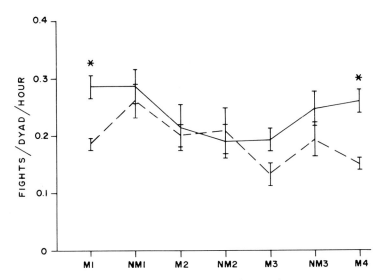

Fig. 14. Female-female fights/dyad/hour. Mean ± SE. Control females (all) _____, experimental females (all) - - - - - - -, significant inter-group difference shown by *.

tially confirmed by an analysis of fights/female/h which revealed a significantly higher overall mean for the controls than for the experimentals (1.04 ± 0.05 vs. 0.78 ± 0.05; F = 12.03, df 1/90, p < 0.001). A seasonal analysis indicated that the overall difference was shaped primarily by mating season behaviors (fig. 15). The controls significantly exceeded the experimentals across the summed mating seasons (F = 12.56, df 1/50, p < 0.001), but not across the summed non-mating seasons.

A second measure of female aggressiveness, number of victories/subordinate/h, was calculated from ad libitum data. This index showed the control females to have been more aggressive than the experimental animals, but also demonstrated that the difference was very small. The control females showed an overall rate of 0.11 ± 0.004 victories/subordinate/h, compared to the experimental females' rate of 0.10 ± 0.004 (F = 5.30, df 1/92, p < 0.05). A seasonal analysis revealed no significant differences during individual reproductive seasons or across the summed non-mating seasons, although the two female populations did differ significantly across the summed mating seasons (F = 4.30, df 1/52, p < 0.05).

The inter-group differences in female aggressiveness and agonistic behavior apparently were not related to the presence of cycling female 370 within the experimental group. 370's

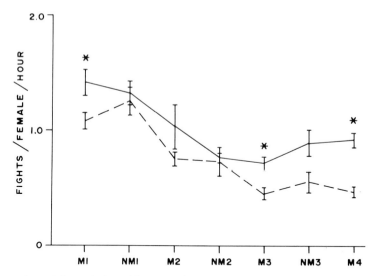

Fig. 15. Fights/female/hour. Mean ± SE. Control females
(all) _____, experimental females (all) - - - - - - -,
significant inter-group difference shown by *.

agonistic behavior did not appear to differ in any significant
way from that of her ovariectomized groupmates. Specifically
with regard to aggressiveness, 370's range of monthly means for
victories/subordinate/h (0.04-0.19) fell well within the range
of means for all other experimental females (0.00-0.24). As
a consequence of her normalcy, 370's agonism data were lumped
with those from the other experimental females.

Heterosexual fights/dyad/h
    The experimentals exceeded the controls in male-female
fights/dyad/h, while the reverse was true for female-male
fights. This combination of aggressive patterns produced nearly equal rates of summed heterosexual fighting. Mean numbers
of heterosexual fights/dyad/h did not differ significantly between the groups across the entire scan sampling period, across
the summed mating seasons, or across the summed non-mating seasons. Seasonal analysis revealed significantly more heterosexual fighting among the experimentals than among the controls
during mating seasons 3 and 4 (F = 5.33, df 1/12, p < 0.05,
and F = 20.46, df 1/12, p < 0.001, respectively). These differences were due primarily to the aggressiveness of the castrated males (fig. 11).

Isosexual fights/dyad/h
In a manner similar to that just described for heterosexual agonism, male-male fighting and female-female fighting combined to produce non-significant inter-group differences in total isosexual fights/dyad/h. This index showed no significant differences at any level of analysis (total post-operative, summed seasons, individual seasons).

Results: Post-operative Agonism (Ad Libitum Data)

Intra-group Differences in Aggressiveness
Using ad libitum data, the males and females of each group were tested for inter-gender differences in aggressiveness. The castrated males were found to be significantly more aggressive (i.e. show more victories/subordinate/h) than the ovariectomized females ($F = 4.05$, df 1/90, $p < 0.05$). Within the control group, females were found to be the more aggressive sex ($F = 22.04$, df 1/90, $p < 0.001$).

Wounds
Information about wounds suffered by the monkeys was routinely collected during ad libitum sampling. Wound data gathered between July 1972 and May 1976 were analyzed to determine whether there were inter- and/or intra-group differences in the severity of post-operative aggressive behavior. For analysis, all wounds were treated equally and counted only once, regardless of size or effect on the victim. Mashed digits were counted as wounds, but limping monkeys whose injury could not be further specified were not counted as wounded. Monthly totals of wounds/male, wounds/female, and wounds/monkey (males and females combined) were calculated for each group. Inter- and intra-group comparisons were conducted using "t" tests.

Across the sampling period, the controls suffered significantly more wounds than the experimentals (0.20 ± 0.02 vs. 0.11 ± 0.01 wounds/monkey/month, $p < 0.001$). Control females were the most frequently wounded sex-class (0.37 ± 0.10 wounds/female/month), suffering significantly more wounds than the control males (0.13 ± 0.02 wounds/male/month, $p < 0.05$), experimental females (0.14 ± 0.02 wounds/female/month, $p < 0.05$), and castrated males (0.08 ± 0.01 wounds/male/month, $p < 0.01$). Castrated males suffered significantly fewer wounds than either their female groupmates or the control males ($p < 0.02$ and $p < 0.05$, respectively). The control males and experimental females did not differ significantly in wound rate.

When data on wounds were summarized across the summed mating and non-mating seasons, it was found that neither the control males nor females showed seasonal changes (table VII). In contrast, the castrated males appeared to suffer more wounding during mating seasons, while the experimental females were wounded most often during non-mating seasons.

Table VII. Wounds/monkey/month

Animals	Summed mating seasons	Summed non-mating seasons
Control males	0.13 + 0.03*	0.13 + 0.03
Control females	0.28 + 0.04	0.29 + 0.07
Castrated males	0.09 + 0.02	0.06 + 0.02
Experimental females	0.11 + 0.02	0.19 + 0.04

* Mean + SE

Dominance
    As reported by LOY and LOY [100], the 33 subjects brought
a well-established dominance network to the study. Hierarchies
within the two age-classes of juveniles approached linearity,
and within-age-class dominance relationships were 95.6% pre-
dictable from knowledge of the juveniles' mothers' ranks within
Cayo Santiago Group A.
    The monkeys' dominance relationships changed somewhat
after group formation and the gonadectomies. In order to per-
form inter-group comparisons of dominance stability, we quantified dominance shifts within each group using the Dominance
Change Score (DCS). Each month the observed dominance relationships (both within and between age-classes) in each group
were compared to the relationships noted during the previous
month. If a relationship had reversed completely (i.e. if "A
always dominant to B" had changed to "B always dominant to A"),
it contributed a full point toward the DCS. Clear relationships which had become ambiguous (i.e. if "A always dominant to
B" had changed to "A and B each defeat the other at least
once") contributed a half point toward the index. Previously
ambiguous relationships which had apparently stablized likewise
contributed a half point. All dominance changes within a group
were totaled, and that sum was divided by the number of dyads
within the group. Since the result of these calculations was
usually a very small number, it was multiplied by 100 to produce
the DCS.
    Between 26 January-21 March 1972 (pre-operative, but with
the groups separated), both controls and experimentals showed
fairly high DCS values (4.58 and 6.25, respectively), indicating shifting relationships in the two groups. The experimentals showed significantly higher DCS values (i.e. less stable
dominance networks) than the controls early in the post-operative period (fig. 16). By non-mating season 2, however, the
two groups showed very similar DCS levels and continued to do
so until the end of the study.

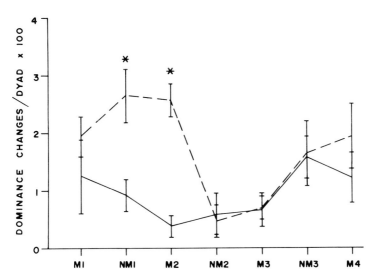

Fig. 16. Dominance change scores. Mean ± SE. Controls (all) _____, experimentals (all) - - - - - - -, significant inter-group difference shown by *.

These findings suggest that dominance instability within the experimental group was initially related to group formation but later was exacerbated by the gonadectomies. Whether the gonadectomy effects were due to changes in physiology or to some other factor is unknown. The operation schedule occasionally resulted in healthy animals being housed with monkeys who were vulnerable to attack due to recent surgery; therefore, a "design effect" cannot be ruled out.

The experimentals' dominance changes were not the result of across-the-board rank increases or decreases by either males or females. As shown in figure 17a, while the average rank for castrated males remained somewhat above the experimental females' mean rank throughout the post-operative period, neither sex showed a strong tendency to rise or fall in rank. Within the control group, males showed a steady post-operative decline in mean rank, while females rose in rank (fig. 17b). As a consequence, the substantial inter-gender difference in mean rank shown by the controls early in the post-operative period had disappeared by the end of the study.

Post-operative rank changes involved both within-age-class and inter-age-class relationships. The experimentals' within-age-class relationships, highly correlated with their mothers' Cayo Santiago ranks pre-operatively, showed a strong post-operative drop in correlation with maternal ranks (fig. 18a). In contrast, the control monkeys showed little change across the study in the correlation between their within-age-class dominance relationships and their mothers' Group A ranks (fig. 18b).

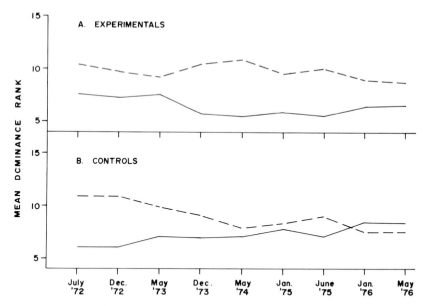

Fig. 17. Dominance rank and sex. Males _____ , females - - - - - - -. Note that within each group large numbers
indicate low rank and small numbers indicate high rank.

Results: Correlations Between Behavioral, Physiological, and Somatic Variables

Males

Tests for relationships between aggressive behavior, testosterone levels, and body size were conducted using a series
of Pearson product-moment correlations. For each of the seven
bleeding dates, six variables were cross-correlated for each
male population:

1. number of subordinates (data from ad libitum samples)
2. victories/h (from ad libitum samples)
3. victories/subordinate/h (from ad libitum samples)
4. plasma testosterone
5. body weight
6. crown-rump length.

As noted earlier, the number of other monkeys that an animal always or occasionally defeated was used to calculate
dominance rank. As expected, within each male population,
number of subordinates was strongly and positively correlated
with number of victories/h. This suggests a relationship between dominance rank and aggressiveness (table VIII). When
aggressiveness was measured in terms of victories/subordinate/
h, however, no clear correlation with number of subordinates

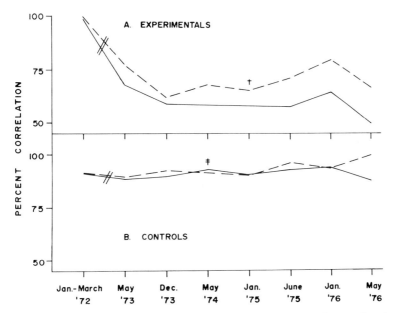

Fig. 18. Percent of observed within-age-class dominance relationships correlated with mothers' (Cayo Santiago) ranks. Males _____, females - - - - - - - -.† February 1975 data used. ‡ June 1974 data used. Crossed lines indicate the long interval between the pre-operative data (January-March 1972) and the first post-operative measurement (May 1973).

was found. Number of subordinates was not significantly correlated with testosterone in either male population (with the exception of the castrates in June 1975, a correlation probably due to chance). Similarly, neither intact nor castrated males showed consistent, significant correlations between number of subordinates and either body weight or crown-rump length (see BERNSTEIN and DRAPER [16] for similar results).

The male populations showed no clear relationships between testosterone and either measure of aggressiveness (table IX). The controls showed one significant and positive correlation with victories/h, but this occurred during the low-testosterone non-mating season and could have been due to chance. Surprisingly, the castrated males presented more evidence than the controls for a link between testosterone and aggressiveness, showing two significant and positive correlations with victories/h and one with victories/subordinate/h. The male populations showed no clear relationships between testosterone and either weight or crown-rump length (table IX).

The two measures of aggressiveness (victories/h and victories/subordinate/h) were strongly and positively correlated

Table VIII. Pearson r values for males

Variables	Groups	Bleeding dates						
		May'73	Dec.'73	May'74+	Jan.'75	June'75	Jan.'76	May'76
Number of subordinates vs. victories per h	Cont.	.6714*	.7459*	.8126*	.8338**	.6813*	.8355**	.9513***
	Cast.	.9111**	.8976***	.7513*	.7374*	.8715***	.9236***	.9556***
Number of subordinates vs. victories per subordinate per h	Cont.	.1693	.5648	.5175	.6658*	.2820	.2956	.7797*
	Cast.	.4724	.3305	-.2344	-.0364	-.1421	.6706*	-.7525*
Number of subordinates vs. testosterone	Cont.	.6071	.3494	.6647	-.4492	-.1474	.2833	.5282
	Cast.	.4227	.5530	.0839	.2492	.7260*	.4976	.1509
Number of subordinates vs. weight	Cont.	.6313	.3231	.0857	-.5861	-.2521	-.5018	-.4797
	Cast.	.3393	.2519	.0627	.2307	.1258	-.4685	-.1899
Number of subordinates vs. crown-rump length	Cont.	.4897	.2615	.2907	-.2181	-.5388	-.3984	-.7724*
	Cast.	.3930	.2908	.1564	-.5682	.3475	-.2120	.1873

+ Number of control males = 7
* $p < 0.05$
** $p < 0.01$

Table IX. Pearson r values for males

Variables	Groups	Bleeding dates May'73	Dec.'73	May'74+	Jan.'75	June'75	Jan.'76	May'76
Testosterone vs. victories per h	Cont.	.0593	.3045	.3945	-.3779	.1128	.2933	.6887*
	Cast.	.1099	.4888	-.0312	.1111	.7176*	.7134*	-.0211
Testosterone vs. victories per subordinate per h	Cont.	-.3714	-.0539	-.0306	-.5368	.4930	-.0098	.3923
	Cast.	-.2264	.2634	-.0952	.1352	-.0754	.7398*	-.4338
Testosterone vs. weight	Cont.	.7170*	.0369	.5296	-.0930	.4347	-.4527	.3222
	Cast.	.0468	.3165	.1577	-.0325	-.1663	.0425	.0549
Testosterone vs. crown-rump length	Cont.	.6587	.0494	.6955	.0651	.7004*	-.2637	-.3122
	Cast.	.0218	.3810	.0805	-.0524	-.0566	.4245	.2756

+ Number of control males = 7

* $p < 0.05$

Table X. Pearson r values for males

Variables	Groups	Bleeding dates						
		May'73	Dec.'73	May'74[+]	Jan.'75	June'75	Jan.'76	May'76
Victories per h vs. victories per subordinate per h	Cont.	.7569*	.8968***	.8698**	.9272**	.8207**	.7570*	.8265**
	Cast.	.7353*	.6683*	.4377	.5865	.3173	.8292***	-.5622
Victories per h vs. weight	Cont.	.0581	-.1791	-.0656	-.3853	-.0479	-.5929	-.3010
	Cast.	.4363	.3104	.0589	-.0106	.1036	-.3521	-.1859
Victories per h vs. crown-rump length	Cont.	-.0298	-.2460	-.1468	.2828	-.4231	-.0101	-.7417*
	Cast.	.4511	.4309	.1813	-.6108	.2126	.0342	.0798
Victories per subordinate per h vs. weight	Cont.	-.2229	-.1056	-.1652	-.1061	.1716	-.3723	-.3088
	Cast.	.1644	.4728	-.1547	-.0803	-.1535	-.1030	-.0734
Victories per subordinate per h vs. crown-rump length	Cont.	-.2912	-.2276	-.4779	.4217	-.1207	.4841	-.4436
	Cast.	.0648	.6264	-.0628	-.3039	-.1953	.2069	.5452

+ Number of control males = 7
* $p < 0.05$
** $p < 0.01$

with each other among the control males. They were apparently, although less strongly, related among the castrated males (table X). Neither index of aggressiveness was related to weight or crown-rump length for either group of males (table X).

Females

Fewer Pearson correlations were calculated for females than for males. Due to the confounding effects of monthly hormonal cycles and pregnancies among the controls, no correlations involving estradiol or progesterone values, or body weight were done.

Within both female populations, number of subordinates was strongly and positively correlated with victories/h (table XI). The same was not true, however, for victories/subordinate/h. This variable was not correlated with number of subordinates among the control females, although among the experimentals there may have been a weak relationship between the two measures. Neither female population showed a relationship between number of subordinates and crown-rump length.

Both the intact and experimental females showed a moderate-to-strong relationship between victories/h and victories/subordinate/h (table XI). Neither measure of aggressiveness was strongly correlated with crown-rump length in either female group.

Discussion

Although they began the study with equal potential for aggressive behavior, the two male populations showed unequal post-operative declines in aggressiveness, with the result that the experimental males were significantly more aggressive than the controls by the end of the investigation (fig. 10, 11, 12, 13). These results were surprising, since numerous studies have reported that among rhesus males aggressiveness is enhanced by testosterone [33, 45, 59, 118, 180] and, as demonstrated, the intact males had much higher testosterone levels than the castrates. A careful reading of the literature, however, reveals a lack of agreement on the effects of castration and testosterone on male aggressiveness. WILSON and VESSEY [181] reported normal frequencies of aggressive behavior for some free-ranging adult male castrates, and BIELERT [19] presented data on lab males whose neo-natal castrations did not affect their aggressiveness between 1-4 years of age. In the same report, BIELERT noted that in pair-tests his castrated males were _more_ aggressive than intact males, and that under some circumstances injecting castrates with testosterone propionate produced a _decrease_ in their levels of aggression. GORDON, ROSE, GRADY and BERNSTEIN [60] reported that increasing testosterone levels through regular injections of human chorionic gonadotropin did not produce an increase in aggression for three of four intact rhesus males living in an all-male group.

Table XI. Pearson r values for females

Variables	Groups	Bleeding dates						
		May'73	Dec.'73	May'74	Jan.'75	June'75	Jan.'76	May'76
Number of subordinates vs. victories per h	Cont.	.9752**	.9948**	.8463**	.8325*	.6569	.9680**	.8369**
	Exper.	.6248	.9902**	.8280*	.9591**	.9340**	.8968**	.8908**
Number of subordinates vs. victories per subordinate per h	Cont.	.5030	.4323	.3107	.5849	.2011	.4428	.1414
	Exper.	-.3634	.8781**	.4088	.0664	.6324	.7954*	.6060
Number of subordinates vs. crown-rump length	Cont.	.0116	-.4839	-.7803*	-.4418	-.6626	-.3240	-.5344
	Exper.	.5269	.4133	-.1385	-.7142	-.6472	-.1393	-.1385
Victories per h vs. victories per subordinate per h	Cont.	.5913	.4715	.7413*	.8901**	.8350*	.5858	.5230
	Exper.	.4201	.8339**	.7356*	.2765	.7929*	.9399**	.8171*

* $p < 0.05$
** $p < 0.01$

Table XI. (continued)

Variables	Groups	Bleeding dates						
		May'73	Dec.'73	May'74	Jan.'75	June'75	Jan.'76	May'76
Victories per h vs. crown-rump length	Cont.	-.1069	-.5456	-.8757*	-.5410	-.3387	-.1371	-.3073
	Exper.	.2957	.3284	.0739	-.7285	-.5665	-.4001	-.4970
Victories per subordinate per h vs. crown-rump length	Cont.	-.3305	-.8837*	-.6493	-.2094	-.1312	.2515	.0349
	Exper.	-.2861	.4714	.3427	.1521	-.0496	-.3141	-.4420

* p < 0.01

Finally, in a recent laboratory study, MICHAEL and ZUMPE [119] found that "...annual changes in plasma androgens were neither necessary nor sufficient to cause...changes in (rhesus male) aggression..." (p. 154).

The relationship between androgens and aggression has recently been reviewed by DIXSON [43]. DIXSON concluded that while androgen (testosterone) apparently enhances the probability that primate males will behave aggressively, "...social factors appear to be more important than androgen in controlling aggressive interactions in groups of monkeys" (p. 51). Following this suggestion, we examined our data for evidence of social variables which could have stimulated aggression among the castrated males.

Males in both groups experienced a post-pubertal (for the castrated males, post-operative) drop in aggressiveness (fig. 12). By itself, this was not unusual. Rhesus males commonly experience a decrease in aggressiveness after the peak values of the infantile/juvenile period [19, 64, 140, 147]. It was surprising, however, for the controls to show a significantly greater decline in aggressiveness than the castrates. Several variables, including pre-natal hormonal stimulation [63, 64, 127, 128] and crowding [3, 155], have been linked to aggressiveness in macaques. Neither of these variables appeared to be important in the present case; they should have had similar effects on the populations of males. The most convincing explanation for the difference in loss of male aggressiveness is that it was a result of the gonadectomies and the interaction patterns they produced (more accurately, helped to maintain) within the experimental group.

Juvenile rhesus monkeys of both sexes tend to be somewhat isosexually oriented with regard to social partners other than their mother.* That is, juvenile males tend to select males more often than females as partners for social activities (especially play), while juvenile females prefer female partners for the same activities [73, 100, 147, 160]. Our control and experimental males started with about equal levels of isosexual orientation and interaction (data to follow; see especially chapter X). As the controls passed through puberty (complete with the beginning of adult testosterone production) and matured, their social orientation shifted from strong isosexuality toward heterosexuality with seasonal isosexual swings. In contrast, the castrated males maintained high levels of isosexual partner preference throughout the post-operative period. It appears that the frequent aggression among the experimental males was at least partially related to this abnormal continuation of frequent male-male interaction. Castration may have prolonged the isosexual association of those experimental monkeys who were inherently most prone to aggression (due to a

---

* The concept of orientation will be explained more fully in chapter X. It is sufficient here to define orientation as an attraction to, and preference for, certain partners.

sex-specific pattern of brain differentiation [63, 64, 127, 128]). This produced significant inter-group differences in male-male fighting (fig. 10). The significantly higher rates of male-female aggression among the experimentals (fig. 11) may well have been due to the isosexually-oriented castrated males' failure to develop the "friendly" bonds with females thought to be formed by normal adult rhesus males [98].

The factors responsible for normal males' pubertal shift toward heterosexual orientation are unclear. Initial sexual interactions and prolonged exposure to mature, sexually attractive females may be important (although the presence of cycling female 370 among the experimental males did not seem to alter their behavior significantly). In addition, it is possible that increased testosterone levels directly affect orientation in some way. If this is true, then ironically the lack of testosterone among the castrated males might have produced their high levels of aggression. Whatever the explanation for the aggressiveness of the gonadectomized males, their behavior demonstrates that high levels of testosterone are not necessary for the maintenance of frequent aggression by chronologically mature rhesus males.

The failure of the control males to show seasonal fluctuations in aggressiveness (fig. 10, 11, 12, 13) was unexpected, since several earlier studies had reported peaks in male aggressiveness coincident with the mating season and/or seasonal peaks in testosterone [45, 59, 86, 118]. The reasons for the control males' lack of seasonal cycles in aggressiveness are unclear.

The control females were more aggressive than the experimental females, especially during the mating seasons (fig. 14, 15). Although the exact explanation for this difference is unknown, it may have been related to differences in hormonal stimulation. TRIMBLE and HERBERT [163] reported that testosterone treatment can increase the aggressiveness of ovariectomized rhesus females. It is also known that intact females experience a mid-cycle peak in testosterone [77]. In addition, estrogen (perhaps acting synergistically with progesterone) may stimulate female aggressiveness [109]. Free-ranging rhesus females have been described as showing increased aggression just before and during the mating season [45], with aggressiveness being especially marked during estrus [24]. During copulation, a female may threaten groupmates in order to incite her sexual partner to mount [93, 117, 186, 187; J. LOY, unpub. obs.]. In addition, it appeared as if the control females of the present study occasionally chased proceptive female groupmates away from particular males. Such inter-female competition for males could have helped produce a high female aggression rate. Whatever the bases for their aggressiveness, the finding that the control females were more aggressive than their male groupmates was quite unusual. Data from free-ranging rhesus adults indicate that the opposite condition is generally true [45].

The data on wounding indicated that the control monkeys engaged in more severe fighting than the experimentals. This finding could have been related to the fact that the control males were heavier and had larger adult canines than the castrated males; therefore, they may have been more likely to wound their opponents during fights. Thus, even though the intact males engaged in fewer fights among themselves and defeated females less frequently than did the castrated males, they might have inflicted more wounds than the latter animals. On the other hand, rhesus females are quite capable of wounding an opponent. Therefore, the high rate of inter-female fighting within the control group (fig. 14) could have contributed significantly to the high wounding rate. Unfortunately, further clarification of this problem is impossible since the actual wounding of monkeys was rarely seen.

Several studies of dominance among nonhuman primates have implicated the gonadal hormones, and especially testosterone, as substances that determine or maintain rank. ROSE, HOLADAY, and BERNSTEIN [142] found a significant positive correlation between dominance rank and plasma testosterone within a captive all-male rhesus group. Among free-ranging rhesus monkeys, some castrated males have been reported to show gradual post-operative decreases in dominance rank [181]. Testosterone therapy has been reported to induce rank increases in intact, pre-pubertal rhesus females [83], gonadectomized rhesus males and females [19, 33], and gonadectomized male and female chimpanzees (Pan troglodytes [21, 32]). While the effects of estrogens on dominance rank have not been as well studied as those of testosterone, estrogen therapy has been found to induce rank increases in ovariectomized chimpanzees [21], but to adversely affect the rank of castrated chimpanzee males [32]. Citing the above studies, as well as several on humans, MAZUR [107] recently developed a general theory suggesting that among primates, testosterone is related to the activities involved in achieving and/or maintaining status.

There are, of course, numerous variables besides testosterone that can affect dominance rank: social relationships with groupmates [174]; agonistic support networks based on kinship or other affectional ties [99, 103, 148, 181]; and "social tradition" based on the history and duration of contact between animals [33, 156]. As a consequence, some hormonal manipulation or therapy studies have failed to produce rank changes [60, 120], numerous studies on stable social groups have failed to find significant correlations between testosterone and dominance rank (rhesus [59]; Macaca fuscata [52]), and one study has even reported the rise of an agonadal rhesus male to the alpha position within a heterosexual group which included four intact adult males [17].

Although care must be taken when interpreting the testosterone data from the present study (since the values represent neither seasonal peaks nor means, and might have been affected by the stress of capture), it appears that the control males provide another example of a lack of correlation between testosterone and either dominance rank or aggressiveness in a stable group of intact rhesus monkeys (tables VIII, IX). Undoubtedly, the lack of a correlation between rank and testosterone was at least partially related to the feral origin of the group [51] and to the fact that early effects of maternal and matrilineal rank continued to influence the controls' dominance network throughout the study (fig. 18b).

Our experimental monkeys maintained a network of frequently reinforced dominance relationships throughout the post-operative period. Although initially less stable than the controls' dominance network (a period of instability that was somehow related to a reduction in the influence of maternal rank on dominance [fig. 16, 18a]), the experimentals' rank relations were as stable as those of the intact monkeys during the last two years of the investigation. These data, plus the lack of correlation between the castrated males' low testosterone values and dominance ranks, suggest that rhesus monkeys do not require hormonal support for the maintenance of clear, relatively stable dominance ranks.

ROSE, HOLADAY, and BERNSTEIN's [142] early study of rhesus hormone-behavior relationships reported a significant correlation between frequency of aggression and testosterone level within an all-male group. Later studies of macaques living in stable heterosexual groups, however, indicated no correlation between male aggression and testosterone (e.g. M. fuscata [52]). The results of the present study provide some support for the conclusion that individual differences in aggressiveness among intact rhesus males are not related to differing testosterone levels (table IX), and suggest that this may be true for castrated males as well.

Studies which define dominance relations in terms of winning/losing fights stand a good chance of finding a correlation between rank and frequency of aggression [52, 142]. The present study was no exception, as all four sex-by-group sub-populations showed significant correlations between number of subordinates and victories/h (tables VIII, XI). Such correlations, however, are based on circular reasoning and measurement. That is, a high-ranking individual must have defeated each of his/her numerous subordinates at least once in order to have demonstrated high rank, and in so doing the animal has also accumulated a high aggression score. Using this procedure, animals who have numerous subordinates, but who defeat each subordinate only rarely, are labeled as "aggressive" individuals. Rhesus monkeys express their aggressiveness within the context of their existing dominance and social relationships, and a monkey is always limited with regard to its number of targets for aggression. (It is true that dyadic relationships may be changed due to repeated attacks by subordinates, but generally rhesus monkeys aggress against animals they are sure to defeat.) Therefore, an index which measures aggression relative to the available targets would seem to provide a more valid measure of "aggressiveness" than simply the number of observed threats, attacks, or victories per hour. For these reasons, we considered victories/subordinate/h to be our best measure of aggressiveness. It was quite interesting to find

that while victories/h was strongly correlated with dominance
rank, number of victories/subordinate/h was not (tables VIII,
XI). Therefore, depending on the measure of aggressiveness,
the present study did or did not document a correlation with
dominance.

Finally, body size (weight and crown-rump length for
males, crown-rump length for females) was not correlated with
dominance or aggressiveness for any sex-by-group sub-population
(tables VIII, X, XI). Similarly, testosterone was not strongly
related to either size variable for either male population
(table IX). (Note, however, that this conclusion might have
been different if testosterone values had reflected either seasonal peaks or means for each male rather than randomly collected single hormone concentrations.)

Summary

The control and experimental monkeys began the study with
equal potential for aggressive behavior and with very similar
dominance networks. During the post-operative period, however,
significant inter-group differences developed for both vari-
bles. Although total post-operative fighting (fights/dyad/h)
was about the same in both groups, the controls engaged in more
severe fighting, as shown by their higher wounding rate. Interestingly, the high rate of wounding among the controls was
not correlated with high male aggressiveness. The control
males were less aggressive than their female groupmates and
also failed to show typical seasonal cycles in aggression. The
control females were more aggressive than the experimental females, perhaps because of high gonadal hormone levels during
the mating seasons. The controls' dominance network was
strongly affected by maternal ranks. It was very stable
throughout the study, although sufficient change occurred to
equalize mean male and female ranks.

The castrated males did not show as great a post-operative
decline in aggressiveness as the control males. Therefore,
they proved to be the significantly more aggressive male population. This difference in male aggressiveness may have been
related to the maintenance of an isosexual orientation among
the experimental monkeys. For reasons that were not completely clear, the experimental monkeys experienced a period of instability in their dominance network early in the post-operative period. During this time of instability, the influence
of maternal rank declined strongly. The experimental monkeys'
dominance network stabilized during the last two years of the
study.

V. Mounts and Copulations

Definitions and Methods

In 1932, ZUCKERMAN [184] theorized that sexual attraction and interactions account for the permanent heterosexual societies of monkeys and apes. During the half century since its proposal, this theory has been attacked by some workers [90] and supported by others [85, 98]. A primary objective of the present study was to explore ZUCKERMAN's theory by creating a "sexless" society (i.e. one devoid of heterosexual attraction) for structural analysis and comparison with a normal monkey group. Toward that end, we assigned sexually inexperienced monkeys to the experimental group and then gonadectomized them before or near puberty in hopes of precluding the development of normal patterns of sexual attraction. This chapter reports the effects of our manipulations on primary sexual behaviors, including mounting and copulating.

The definitions pertaining to sexual patterns were as follows:

1. Mount: The mounter contacted the mountee from the rear, usually grasping the mountee's waist manually and ankles/ calves with one or both feet (fig. 19a, b) [111, 175]. Intromission sometimes occurred during male-to-female mounts, and pelvic thrusting often occurred regardless of the sexes of mounter and mountee. Most mounts were of the double foot-clasp variety, but postures that deviated from this pattern were recorded as mounts if they included dorsal-ventral positioning, manual grasping of the mountee's back or sides, and genital approximation (fig. 19c, d). Atypical mounts were most frequently shown by young monkeys [100, 113] and females [1, 116].

2. Series-mounting: This pattern was defined as two or more mounts, closely spaced temporally and usually connected by the partners grooming, sitting-touching or sitting-close to one another [98, 100]. Mounts connected by play (e.g. wrestling) were not regarded as serial.

3. Copulation: A single mount or series-mounting which ended in intra-vaginal ejaculation was classified as a copulation. Ejaculation was marked by the cessation of male pelvic thrusting and by male body rigor and muscular contractions [24, 98, 111].

4. Vaginal plug: A plug consisted of coagulated ejaculate in a female's vulva. This was regarded as good evidence of recent copulation.

5. Sex skin colors: During ad libitum sampling, the colors of each female's face, perianus, and posterior thighs were scored by eye on a five-point scale (0 = no color, 4 = bright red). The sex skin index consisted of the daily average of scores from the three areas [100].

Fig. 19. Mounting postures observed in the study groups.
A. Control male 266 performs a double foot-clasp mount on female 242.   B. A double foot-clasp mount between castrated
males 246 (mounter) and 301.   C.   A mount between experimental
females 273 (mounter) and 366. Note the lack of foot-clasping.
D. Control female 277 single foot-clasp mounts female 235
(whose infant clings to its mother's belly).

C

Scan sample data allowed accurate quantification of the monkeys' mounting behavior. During scan sampling, all mounts were tallied as individual interactions, whether they occurred singly or serially. Furthermore, no attempt was made to categorize mounts according to the context in which they occurred (e.g. play, agonism, consortship) or the presence or absence of ejaculation. When series-mounting was observed, a special notation of the participants' identities and the number of mounts was made on the checksheet.

Several behavioral indices were derived from the scan sample mounting data. Most of these indices were very similar those used to measure agonistic behavior; therefore details on calculation procedures will only be repeated where necessary. The mounting indices were:

1. Mounts/dyad/h
2. Male-male mounts/male-male dyad/h
3. Male-female mounts/heterosexual dyad/h
4. Mounts/male/h
5. Female-female mounts/female-female dyad/h
6. Female-male mounts/heterosexual dyad/h
7. Mounts/female/h
8. Mounts received/male/h: This index was calculated by finding the total number of times males were mounted by a partner of either sex. That figure was divided first by the number of males in the group, and second by the month's total scan hours.
9. Mounts received/female/h: Calculations the same as those for index 8, only for females.
10. Isosexual mounts/isosexual dyad/h
11. Heterosexual mounts/heterosexual dyad/h

Although series-mountings and copulations occasionally were seen during scan sampling, most of the information on these interactions came from ad libitum observations. These data were used to check for correlations between individuals' dominance ranks or testosterone levels and mating behavior. Significant correlations were considered suggestive of real relationships, despite the possibility of unequal sampling of subjects during ad libitum observations.

Results: General Comments on Mount Posture and Ejaculation

Mounting postures used by controls and experimentals appeared to be essentially the same. In both groups most mounts were of the double foot-clasp variety (fig. 19). This was not surprising. All of the monkeys presumably had experienced a normal infancy on Cayo Santiago, and numerous studies have shown the importance of early social experiences on the development of mounting behavior [65, 175, 176].

With regard to ejaculatory behavior, all control males were seen to ejaculate intra-vaginally at least once during the post-operative period. In contrast, only one experimental male (231) showed intra-vaginal ejaculation, and the one observed incident occurred on 15 March 1972, prior to 231's castration. None of the experimental males showed either seminal emission or the motor patterns of ejaculation during post-operative mounting.

Results: Mounting (Scan Data)

Mounts/dyad/h
Between July 1972-May 1976, the experimental monkeys showed a significantly higher overall mounting rate than the controls (0.068 ± 0.004 vs. 0.049 ± 0.008 mounts/dyad/h; F = 3.986, df 1/90, p < 0.05). The difference was related to the fact that the experimentals' mounting rate remained constant from season to season, while the controls' rate declined sharply during non-mating periods (controls' mounts/dyad/h, summed mating seasons: 0.060 ± 0.014; summed non-mating seasons: 0.036 ± 0.005). Therefore, although the groups differed significantly across the summed non-mating seasons (F = 21.618, df 1/38, p < 0.001), no difference was found for the summed mating periods.

Male-male mounts/dyad/h
Throughout the entire post-operative period, the castrated males showed much higher levels of male-male mounting than the controls. There were significant inter-group differences across the total scan sample period (castrates 0.172 ± 0.012 vs. controls 0.047 ± 0.007; F = 76.014, df 1/90, p < 0.001), across the summed mating seasons (castrates 0.167 ± 0.019 vs. controls 0.040 ± 0.010; F = 36.535, df 1/50, p < 0.001), and across the summed non-mating seasons (castrates 0.178 ± 0.016 vs. controls 0.056 ± 0.011; F = 42.235, df 1/38, p < 0.001). A seasonal analysis revealed significant inter-group differences during six of the seven post-operative seasons (fig. 20).

The control males experienced a steady decline in male-male mounting as they grew older, while the castrated males maintained high frequencies throughout the post-operative period (with the possible exception of mating season 4) (fig. 20). Neither population of males showed seasonal fluctuations in male-male mounting.

Male-female mounts/dyad/h
Surprisingly, there were few inter-group differences in the rate of male-female mounting. The two groups did not differ significantly across the entire post-operative period, across the summed mating seasons, or across the summed non-mating seasons. The controls did exceed the experimentals during mating season 1, but the only other significant seasonal difference (during non-mating season 2) found the experimental monkeys with the higher rate (fig. 21).

The controls displayed clear seasonal cycles in male-female mounting, while the experimentals did not. A mounting peak was expected among the controls during mating season 4,

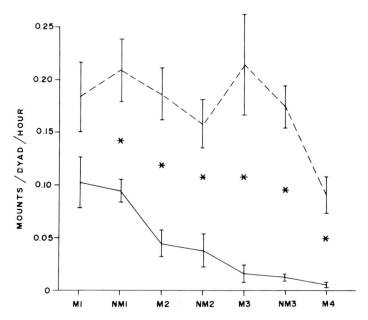

Fig. 20. Male-male mounts/dyad/hour. Mean ± SE. Control males (all) _____, castrated males (all) - - - - - - - -, significant inter-group difference shown by *.

but failed to materialize. The missing peak was probably re-lated to the fact that the two highest-ranking control males monopolized mating behavior during mating season 4, and appar-ently squelched the heterosexual mounting of the other six males (see section on Copulations).

Mounts/male/h
Peaks in male-female mounting among the controls prevented significant inter-group differences in mounts/male/h during mating seasons 1 and 2 (fig. 22), and therefore across the sum-med mating seasons. The castrated males did exceed the con-trols' mounting rate during the remaining five seasons, how-ever. This fact produced significant inter-group differences across the summed non-mating seasons (0.817 ± 0.065 vs. 0.353 ± 0.064 mounts/male/h; F = 26.126, df 1/38, p < 0.001) and across the total post-operative period (0.799 ± 0.051 vs. 0.512 ± 0.101; F = 6.468, df 1/90, p < 0.025). As noted, most of the castrated males' mounting was directed toward other males.

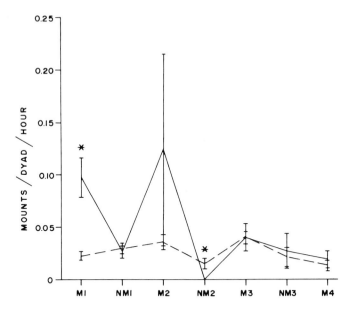

Fig. 21. Male-female mounts/dyad/hour. Mean ± SE (missing SEs were less than 0.001). Controls (all) _____, experimentals (all) - - - - - - -, significant inter-group difference shown by *.

Female-female mounts/dyad/h

Although the two female populations began the post-operative period showing similar rates of female-female mounting, the controls displayed an increase in this behavior as they grew older, while the experimentals showed a decline (fig. 23). Due primarily to a significant difference during mating season 3, the controls showed a higher mounting rate across the summed mating seasons ($F = 9.065$, df 1/50, $p < 0.01$). Although the groups' rates for the summed non-mating seasons were not significantly different, the controls exceeded the experimentals in overall post-operative mounting (0.060 ± 0.010 vs. 0.022 ± 0.003; $F = 12.646$, df 1/90, $p < 0.001$).

Female-male mounts/dyad/h

Female-to-male mounts occurred infrequently in both groups, and their occurrence declined as the females grew older. There were no significant inter-group differences for individual seasons or across the summed non-mating seasons. The experimental females, however, did significantly exceed the con-

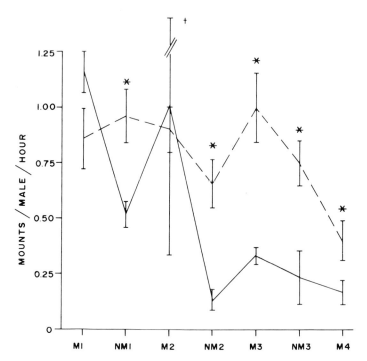

Fig. 22. Mounts/male/hour. Mean ± SE. Control males (all) _____, castrated males (all) - - - - - - -, significant inter-group difference shown by *. † Controls' SE = 0.672.

trols across the summed mating seasons (F = 4.551, df 1/50, p < 0.05) and across the total post-operative period (0.003 ± 0.001 vs. 0.001 ± 0.0004; F = 4.420, df 1/90, p < 0.05).

Mounts/female/h
Due to their higher rate of female-female mounting, the controls showed a significantly higher overall rate of mounts/female/h than the experimentals (0.188 ± 0.030 vs. 0.096 ± 0.016; F = 7.183, df 1/90, p < 0.01). As was true for mounts between females, the only significant seasonal difference in mounts/female/h occurred during mating season 3 (fig. 24). This difference (plus slightly higher control values during mating seasons 2 and 4) did, however, produce a significant inter-group difference across the summed mating seasons (controls 0.208 ± 0.045 vs. experimentals 0.099 ± 0.023; F = 4.677, df 1/50, p < 0.05). The two groups did not differ in mounts/female/h across the summed non-mating seasons.

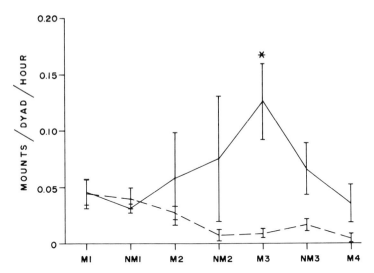

Fig. 23. Female-female mounts/dyad/hour. Mean ± SE.
Control females (all) _____, experimental females (all)
- - - - - - - -, significant inter-group difference shown by *.

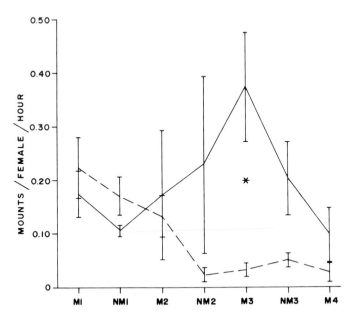

Fig. 24. Mounts/female/hour. Mean ± SE. Control females
(all) _____, experimental females (all) - - - - - - - -,
significant inter-group differences shown by *.

In order to determine the effect of cycling female 370 on the experimentals' mounting behavior, portions of the ad libitum data were analyzed. A stratified random sample of 200 male-to-female mounts, 50 from each of the four post-operative mating seasons, was drawn. The hypothesis that male-female mounting was evenly distributed across the experimental females was tested and rejected (χ^2 = 149.25, df 7, p < 0.001). It was found that three females (370, 366, and 367) participated in more male-female mounting than expected, while four females were mounted much less frequently than expected (260, 272, 273, and 386). Interestingly, all three of the frequently mounted females showed at least one period of post-operative vaginal bleeding, with female 370 showing regular monthly menstrual and sex skin color cycles. These results suggested that 370, along with the other females who apparently experienced some recrudescence of sexual cycling, may have been more attractive, proceptive, and/or receptive (APR) to the castrated males than were completely acyclic females [14, 76, 82, 87]. Any combination of the APR variables could have raised male-female mounting rates for the females in question.

Non-random selection of partners for male-female mounting continued among the experimentals during the non-mating seasons. Analysis of a stratified random sample of 170 mounts from the three non-mating seasons led to the rejection of the hypothesis of an even distribution of mounts across females (χ^2 = 129.24, df 7, p < 0.001). Females 367, 366, and 370 once again were found to have been mounted most frequently. They were joined by female 354, whose high rate of being mounted could have been related to her low rank. Females 260, 272, 273, and 386 were mounted less frequently by males than expected. Interestingly, 370's total mounts did not exceed her expected value, suggesting that she may have been more attractive, proceptive, and/or receptive to males during the mating seasons than during non-mating periods.

As suggested above, dominance rank apparently had a strong influence on the pattern of partner selection for the experimentals' male-female mounts. Female 367, a middle-ranking to low-ranking member of the experimental group, was mounted more frequently by males than was high-ranking female 370, even though 370 was more "normal" (physiologically and somatically). It appeared that many of the castrated males were inhibited from mounting 370 because she out-ranked them. The matter was complicated, however, by the fact that 367 rose in rank over several of her groupmates midway through the post-operative period. Her rank changes were accompanied by frequent fighting. It was common for her to seek support from males during a fight and, in the process, to be mounted. Therefore, many of the mounts that 367 received may have been related to agonism rather than to sex.

During both mating and non-mating seasons, female 370 mounted males much more often than did any of the other experimental females. 370 also mounted her female groupmates frequently. Both types of mount may have been stimulated by 370's hormonal cycles [1, 13, 25, 116], but her high rank certainly gave her the freedom to be a frequent mounter. The factors affecting the mounting behavior of the experimental females were extremely complex, as shown by the fact that three of the ovariectomized animals approached or exceeded 370's rate of female-female mounting. Furthermore, for reasons unknown, 366 and 367 were the most frequent recipients of female mounts, while 370 (perhaps due to her rank) was rarely mounted by females.

Mounts received/male/h

The castrated males surpassed the control males in the total frequency of mounts received (regardless of mounter's sex). They showed significantly higher rates across the entire post-operative period (0.630 $\pm$ 0.045 vs. 0.168 $\pm$ 0.026; F = 79.524, df 1/90, p < 0.001), across the summed mating seasons (0.623 $\pm$ 0.068 vs. 0.145 $\pm$ 0.035; F = 39.674, df 1/50, p < 0.001), and across the summed non-mating seasons (0.640 $\pm$ 0.057 vs. 0.199 $\pm$ 0.038; F = 41.651, df 1/38, p < 0.001). A seasonal analysis indicated significantly higher values for the castrates during all seven post-operative seasons.

Mounts received/female/h

In contrast to the mounts-received data for males, it was the control females who showed the highest rate of mounts received/female/h. Across the post-operative period, the control females were mounted 0.562 $\pm$ 0.120 times/female/h, compared to the experimentals' rate of 0.295 $\pm$ 0.030 (F = 4.673, df 1/90, p < 0.05). Significant inter-group differences during mating season 1 (F = 9.509, df 1/10, p < 0.025; due primarily to male-female mounts, see fig. 21) and mating season 3 (F = 7.316, df 1/12, p < 0.025; due primarily to female-female mounts, see fig. 23) combined to produce a significantly higher rate for the controls across the summed mating seasons (0.742 $\pm$ 0.202 vs. 0.294 $\pm$ 0.042; F = 4.729, df 1/50, p < 0.05). The summed non-mating season values did not differ significantly between the two female populations.

Isosexual mounts/dyad/h

Primarily because of frequent male-male mounting among the castrates, the experimentals significantly exceeded the controls in total isosexual mounting across the entire scan sampling period (0.112 $\pm$ 0.010 vs. 0.052 $\pm$ 0.006; F = 28.466, df 1/90, p < 0.001), across the summed mating seasons (F = 11.747, df 1/50, p < 0.01), and across the summed non-mating seasons (F = 25.098, df 1/38, p < 0.001). A seasonal analysis revealed significantly higher values for the experimentals during four post-operative seasons.

Heterosexual mounts/dyad/h

The two groups differed significantly in overall rates of heterosexual mounting only during mating season 1, when the controls showed high frequencies of male-female mounting (fig.

21). As a consequence, there were non-significant inter-group differences in heterosexual mounts/dyad/h across the entire post-operative period, as well as across the summed mating and summed non-mating seasons.

Results: Series-mounting and Copulation (Ad Libitum Data)

Male-female series-mounting

Although during group formation most of the sexually active animals were assigned to the control group, pre-operative observations indicated a fair balance between the groups with regard to numbers of males seen to series-mount a female. Five of the eight control males were observed to series-mount a female during the pre-operative period, although none was seen to ejaculate intra-vaginally. Among the experimentals, four (of nine) males series-mounted females pre-operatively, and one series (by male 231) ended with ejaculation.

As expected, the gonadectomies had a depressing effect on mating behavior [115], and post-operatively the controls engaged in male-female series-mounting much more frequently than the experimental monkeys. Between April 1972 and May 1976, the controls averaged 0.004 ± 0.001 male-female series-mountings/ heterosexual dyad/ad libitum h, compared to the experimentals' rate of 0.001 ± 0.0002 ($F = 11.491$, df 1/98, $p = 0.001$; these figures include both series with and without intra-vaginal ejaculation). In addition, it appeared that male-female series within the control group were considerably longer (averaging over six observed mounts) than those within the experimental group (which averaged two to three observed mounts).

The controls showed clear seasonal cycles in the frequency of male-female series-mounting, with peak values occurring during the mating seasons (fig. 25). Control means for male-female series-mountings significantly exceeded the experimentals' means between April-July 1972 and during mating seasons 1 and 3. They fell to or below the experimentals' values during most non-mating seasons. Although the castrated males' rate of series-mounting females was low throughout the study, it may have shown weak seasonal cycles (similar to those of the control males) after the first non-mating season (fig. 25).

All of the experimental males (with the exception of male 301) were observed to series-mount a female during the post-operative period. The number of male-female series-mountings/ castrated male ranged from 3-26. In contrast to the intact males, all of whom were seen to copulate post-operatively, none of the castrated monkeys showed the motor patterns of ejaculation after gonadectomy [115].

A Kolmogorov-Smirnov one-sample test run on the experimentals' male-female series-mounting data resulted in the rejection of the hypothesis that females were being series-mounted equally ($p < 0.01$). An examination of the data indicated that females 370, 366, and 367 (all of whom showed at

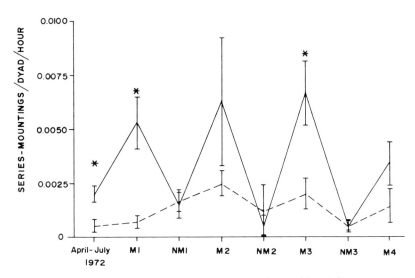

Fig. 25. Male-female series-mountings/dyad/hour. Mean ±
SE. Controls (all) _____, experimentals (all) - - - - -
- - - - - - - -, significant inter-group difference shown by *.

least one instance of vaginal bleeding post-operatively) were
series-mounted relatively frequently, while 260, 273, and 354
were rarely series-mounted by males.

A separate Chi-square test was run on the series-mounting
data from cycling experimental female 370. The result (χ^2 =
5.26, df 1, p < 0.05) indicated that the castrated males mounted her more often than expected by chance. The 20 male-female
series-mountings involving 370 averaged 2.7 (observed) mounts.
Although the series occurred during all stages of her menstrual
cycle, their frequency peaked during the follicular and mid-
cycle stages (fig. 26). Interestingly, the castrated males
series-mounted 370 significantly more often during the mating
seasons than expected (χ^2 = 7.74, df 1, p < 0.005). This may
partially explain the experimentals' weak seasonal cycles in
male-female series-mountings (fig. 25).

Although post-operatively there was no significant correlation between the experimental males' dominance ranks and their
frequencies of series-mounting females, it was clear that for
certain dyads mounting behavior was affected by the partners'
relative ranks. For example, castrated male 241 was not seen
to series-mount 370 while she out-ranked him. Male 241, however, rose in rank over 370 in February 1974. Between then and
the end of the study, he series-mounted her on nine occasions.

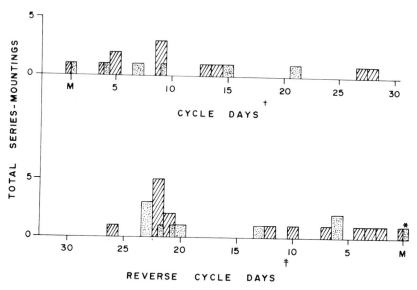

Fig. 26. Heterosexual series-mountings for experimental
female 370. Male-to-female series shown by cross-hatched bars,
female-to-male series shown by stippled bars, menstruation
shown by M. † Forward count starts with the first day of
bleeding. ‡ Reverse count starts on the day prior to the first
day of bleeding. * This female-male series took place on day
two of a three day blood flow.

Copulation
    Copulations occurred only within the control group. Dur-
ing the post-operative period all of the intact males were observed to copulate at least once (range of 1-34 copulations/
male). Dominance rank was an important factor influencing the
control males' sexual behavior, as shown by a significant and
positive Spearman rank correlation coefficient between dominance rank and copulatory frequency across the summed mating
seasons (r_s = 0.759, p < 0.05). The influence of male dominance rank within this confined group was most clearly seen
during the last mating season, when only the top two (of eight)
control males were seen to copulate, and 93.75% of the copulations were by alpha male 266.
    Spearman correlations between the controls' copulatory
frequencies and testosterone values were calculated for mating
seasons 2-4 and for the summed mating seasons. All of these
correlations were non-significant, suggesting (within the limi-
tations of our hormonal data) that the two variables were not
related.
    The copulatory behavior of alpha male 266 deserves special
comment. It may have been somewhat atypical and it may have

Table XII. Observed mounts/copulation for the control males

| Mating season | Male 266 | | All other control males | |
	Number of copulations	Mean mounts per cop- ulation*	Number of copulations	Mean mounts per cop- ulation
1	3	23.0 $\pm$ 8.52	4	6.25 $\pm$ 1.14
2	13	23.46 $\pm$ 5.44	13	7.62 $\pm$ 1.56
3	2	18.0 $\pm$ 8.49	8	3.75 $\pm$ 0.73
4	14	7.64 $\pm$ 1.22	1	2.0 ------

* Plus/minus the standard error.

had a strong effect on certain of the controls' behavioral indices. Male 266 apparently showed unusually long copulatory
series during the first three mating seasons [93, 115]. As
shown in table XII, 266 averaged 18-23.46 observed mounts/cop-
ulation during mating seasons 1-3, while the other control
males were mounting 3.75-7.62 times per copulation. Over 40%
of 266's copulations between August 1972-March 1975 included 30
or more mounts, with the longest comprising 67 observed mounts.
The longest observed series-mounting by 266 occurred on 12
August 1974, when he was seen to mount female 242 a total of
113 times without ejaculating.

For some reason, probably not related to his recent vasectomy [126], 266 was not observed to copulate frequently during mating season 3 (table XII). Although one of his two copulations included 30 mounts, his overall decline in mounting
undoubtedly contributed to the low control mean for male-female
mounts/dyad/h for that season (fig. 21). By the fourth mating
season, however, 266 was not only sexually active once again,
but he clearly monopolized the controls' copulatory behavior.
His mating behavior, however, had changed somewhat, and his
mean number of observed mounts/copulation during the last
mating season was 7.64. It appeared as if 266's monopoly of
the mating behavior within his group (which may have been at
least partly due to his active disruption of other males' sexual interactions) was related to the controls' overall low
levels of male-female mounts and series-mountings/dyad/h during
mating season 4 (fig. 21, 25). Additionally, 266's lower rate
of mounts/copulation might have been partially responsible for
the controls' low number of mounts/male/h during the final
mating season (fig. 22).

Male-male series-mounting

Series-mountings between males were observed in both groups, but were much more common among the castrates than among the controls. Across the total post-operative period, the castrates showed 0.008 ± 0.0007 male-male series-mountings/ male-male dyad/h, compared to the controls' rate of 0.002 ± 0.0003 (F = 63.288, df 1/98, p < 0.001). A seasonal analysis indicated that the controls' rate of male-male series-mounting declined across the study, while that of the castrates was maintained at relatively high levels (fig. 27). The experimental males showed significantly higher frequencies of male-male series-mounting than the controls during six of eight post-operative periods (fig. 27) (data from ad libitum samples; there are eight post-operative periods when April-July 1972 is included).

Within both groups, male-male series-mountings were short, averaging two to three observed mounts. Neither mounter nor mountee was ever observed to ejaculate during a male-male series in either group (total N = 221 series). The castrates consistently showed much higher rates of male-male than male-female series-mounting post-operatively (compare figures 25 and 27). Among the controls, male-female series occurred more frequently during the mating seasons, but fell to or below male-male levels during the non-mating seasons. Neither group showed a seasonal cycle in inter-male series-mounting.

Each of the eight castrated males who survived the entire study was seen to series-mount other males on at least 10 occasions (range of 10-69 male-male series-mountings/male). Although the mean post-operative ranks of the two experimental monkeys (males 246 and 371) who most frequently series-mounted other males were one and three, respectively, there was no general relationship between rank within the male hierarchy and frequency of male-male series (r_s = 0.502, p > 0.05). Nonetheless, the majority (79.6%) of the experimentals' inter-male series-mountings featured the dominant partner as mounter and the subordinate as mountee.

Finally, among the experimental monkeys, there appeared to be little relationship between the males' heterosexual and isosexual series-mounting activities, as a Spearman correlation coefficient between ranked frequencies of male-female and male-male series was non-significant (r_s = 0.303, p > 0.05).

Female-female series-mounting

Inter-female series-mounting occurred significantly more often among the controls than among the experimentals across the entire post-operative period (0.006 ± 0.0011 vs. 0.001 ± 0.0003 series-mountings/female-female dyad/h; F = 17.356, df 1/98, p < 0.001). While series-mountings between females declined post-operatively among the experimental monkeys, the controls showed an increase in this behavior and significantly exceeded the experimentals during the last two mating seasons (fig. 28). Neither group showed clear seasonal cycles in fe-

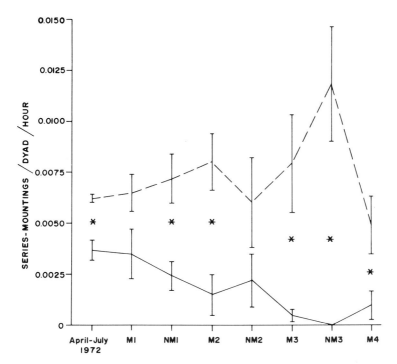

Fig. 27. Male-male series-mountings/dyad/hour. Mean ± SE
(missing SEs were less than 0.0001). Control males (all)
_____, castrated males (all) - - - - - - - -, significant
inter-group difference shown by *.

male-female series, and in both groups the average length of
such series was two to three observed mounts.
    Cycling female 370 participated in only one of the 23 fe-
male-female series-mountings observed post-operatively within
the experimental group. On that occasion 370 mounted female
367 twice in close succession during the luteal phase of 370's
menstrual cycle.

    Female-male series-mounting
    The least frequent form of post-operative series-mounting
was female-to-male. Such series were observed only twice
among the controls (mean observed mounts/series = 2.0) and 15
times among the experimentals (mean observed mounts/series =
2.3). Although an ANOVA indicated a significant inter-group
difference in the overall frequencies of female-male series-
mountings/heterosexual dyad/h (F = 7.384, df 1/98, p < 0.01),
it was clear that the difference was not due to ovariectomy
since cycling female 370 accounted for 13 (86.7%) of the ex-

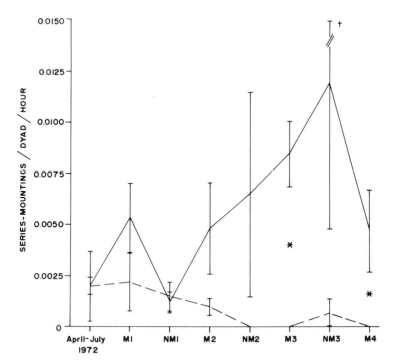

Fig. 28. Female-female series-mountings/dyad/hour. Mean
+ SE. Control females (all) _____, experimental females
(all) - - - - - - -, significant inter-group difference shown
by *. † Controls' SE = 0.0072.

perimentals' F-M series. 370 averaged 2.8 observed mounts/F-M
series. While the mountings occurred throughout her menstrual
cycles, they appeared to be most common during the follicular
and mid-cycle stages (fig. 26) (see MICHAEL, WILSON, and ZUMPE
[116] for similar data from laboratory females).

Discussion

The effects of castration on the mounting and ejaculatory
behaviors of rhesus males are influenced by numerous factors
including social and sexual experiences prior to gonadectomy,
age at castration, and post-operative stimuli. BIELERT's [19]
work showed that some neo-natally castrated rhesus males can
develop the foot-clasp mount posture and eventually display
the ejaculation response if given adequate social experience.
Other studies have reported that normally-reared males castrated as adults may continue to mount females and display the

motor patterns of ejaculation long after their operations [98, 114, 115, 130, 181]. Given these reports, we expected a post-operative continuation of mounting by the La Parguera experimental males. Nonetheless, we were surprised by the frequency of mounting and the males' partner preferences.

In our discussion of agonism and dominance, we suggested that the high rates of aggression shown by the castrated males could have been due to a surgery-related retention of an isosexual orientation pattern. The castrates are described as having retained an attraction to, and preference for, same-sex social partners because we believe such an orientation to be primarily characteristic of juvenile monkeys (evidence for this conclusion is presented in chapter X). This "isosexual orientation theory" can be used to explain not only the castrates' aggressiveness, but their mounting behavior as well.

The experimental males showed much higher rates of mounting than the control males, and most of the castrates' mounts were male-male. Seasonal means for male-male mounting among the experimentals were 5-10 times greater than their rates of male-female mounting. Furthermore, the castrates significantly exceeded the controls in male-male mounts during six (of seven) post-operative seasons (fig. 20). These inter-group differences in mounting may have been related to the fact that the castrated males continued to fight and play with one another at relatively high rates well into chronological adulthood [101] (see chapters IV and IX). Since mounting often occurs in conjunction with fighting and play (rhesus [6, 25, 94, 160]; M. fuscata [68]), frequent participation in both these activities probably contributed to the castrated males' relatively high isosexual mounting rates. In addition, male-male mounts may have been used occasionally as greetings [6, 85] and/or to affirm dominance relationships [6, 25, 85, 100]. Thus the dominance instability experienced by the experimentals early in the post-operative period could also have contributed to their high mounting rates.

Frequent mounts among the castrated males therefore may have been simply by-products of frequent agonism and play. It is important to note that while this explanation implies certain motivations (aggressiveness/submissiveness, playfulness) for most male-male mounts, other motivations are not ruled out. Indeed, it may be impossible to ascertain motivation for many single mounts [68]. It seems reasonable, however, to assume some degree of sexual motivation for male-male series-mountings, given the similarity between these interactions and heterosexual copulations. If this interpretation is correct, then apparently the sexual attraction among the castrated males was much stronger than that found among the control males (fig. 27) or between experimental males and females (compare figures 25 and 27). It appears, therefore, that the "sexless" society we hoped to produce actually contained a high degree of isosexual eroticism, at least among its male members.

The factors responsible for the experimental males' isosexual erotic orientation are not clear. Several variables could have been involved: age at castration, low hormone levels after castration, and a lack of attractive, proceptive, receptive females in the experimental group. The ways in which these variables may have affected the males' behavior are examined below.

The sexual behavior of rhesus monkeys may be viewed as based on four factors: (1) pre-natal brain differentiation, (2) socialization, (3) stimulation by steroidal hormones (most of gonadal origin) during adulthood, and (4) stimulation by conspecifics during adulthood. Experimental manipulation of any of these variables (or several at once) can strongly affect behavior. As noted earlier, since all of our subjects were born and raised (at least through 1.5 years of age) on Cayo Santiago, we assumed that variables 1 and 2 were normal for the species. Typically, the pubertal beginning of adult hormone production by the gonads is coincident with (and presumably stimulates) the beginnings of erotic attraction and erotic interaction between males and females [140]. This heterosexual erotic orientation (and its attendant interactions) may wax and wane seasonally along with the supporting gonadal hormones and conspecific stimulation [58, 165, 167, 168]. If a male is castrated after the development of heterosexual erotic orientation (i.e. post-pubertally), he may continue to be attracted to and interact with (including copulate with) females for long periods post-operatively [98, 114, 130, 181]. If, however, castration is done during the juvenile period of isosexual attraction and interaction, as was true during the present study, males may continue to be isosexually oriented indefinitely and (presumably due to familiarity with frequent social partners [54]) develop unusually strong erotic relationships with male groupmates. According to this model, male gonadal hormones, and especially testosterone, have both an "organizing" and an "activating" role in the development of male sexual orientation and behavior (see GOY and MCEWEN [63], and especially the data summarized therein by DÖRNER). Castration may have precluded the activation of a pubertal shift toward heterosexual eroticism among our experimental males, leaving them in the state of isosexual orientation they had experienced as juveniles.

The castrated males' isosexual erotic orientation, however, may not have been entirely due to their surgeries and low testosterone levels. One cannot refute the suggestion that the castrated males would have been significantly more heterosexually inclined if their group had included more APR females. That the experimental males were capable of heterosexual erotic orientation, despite their low testosterone levels, was evidenced by their continuation of some male-female mounting and series-mounting throughout the study, and by their tendency to series-mount cycling female 370 most often during the mating seasons and the "high estrogen" portions of her menstrual cycles (fig. 26) (see KEVERNE [87] for a discussion of the hormonal correlates of sexual receptivity and attractiveness in rhesus females). In sum, the relative contributions of (1) gonadectomy and altered testosterone levels, and (2) the paucity of APR female partners to be castrated males' isosexual erotic orientation are unclear.

As expected, the control males (after experiencing puberty and adult levels of testosterone) developed a clear seasonal pattern in which heterosexual erotic orientation and interactions alternated with relative isosexuality. Male-female mounting within the control group showed the seasonal cycles characteristic of free-ranging rhesus monkeys (fig. 21, 25) [6, 85, 94, 98], with mounting rates peaking during the periods of maximum testosterone production. The testosterone values for individual control males, however, were not correlated with copulatory frequency [59].

In contrast to testosterone levels, dominance rank among the control males was found to be positively and significantly correlated with copulatory frequency over the total post-operative period. This correlation was probably strengthened by the small size of the corral and its lack of cover. These conditions allowed high-ranking males to monitor (and, if they chose, to disrupt) the sexual activities of subordinates. Positive correlations between male rank and sexual activity have been reported for both free-ranging rhesus monkeys [25, 37, 85, 94, 98] and captive rhesus groups [59], although in some cases the relationships were rather weak. In a study designed to directly measure male reproductive success (by paternity exclusions) within a captive rhesus group, DUVALL, BERNSTEIN, and GORDON [49] found some positive correlation with rank, but the correlation was not consistent over two years of investigation. Similarly, in a more recent paternity study using six rhesus groups, SMITH [153] found a positive association between male rank and reproductive success, but noted that the strength of the association varied among groups and breeding seasons. All of these studies suggest a positive relationship of some degree between male rank and reproductive activity/success among rhesus monkeys, comparable to the relationships reported for other primate species (M. fuscata [71], but see EATON [50, 51] for contradictory results; Papio cynocephalus [75, 125]). It is clear, however, that rank is only one factor affecting a male's sexual behavior. It may also be influenced by (among other things) age, personality, perference for certain females, and female preferences for males (see LANCASTER [89] for a discussion of the numerous variables affecting primate reproductive success).

Dominance rank also affected certain aspects of the mounting behavior of females. Cycling experimental female 370 was able to mount her groupmates freely because of her high rank, and she almost single-handedly produced the experimentals' significantly higher rate of female-male series-mounting. The higher rates of female-female mounting shown by the controls (fig. 23, 28) may well have been related to the fact that these females were repeatedly experiencing normal hormonal cycles and estrus. Several studies have reported that during the follicular stage and during estrus rhesus females frequently mount and/or are mounted by other females [1, 25, 97]. (This is also true for other mammalian species [13].) It is interesting to speculate, therefore, that the peaks in female-female mounting which occurred during the last half of the post-operative period were related to an increase in the number of menstrual cycles and/or time in estrus shown by the control females. Such an increase (as compared to mating seasons 1 and 2) almost certainly occurred, since the control males' vasectomies prevented conceptions after May 1974.

Summary

Despite their low testosterone levels, the castrated males showed more frequent mounting behavior than the male controls. In contrast to the intact males, who showed seasonal peaks in heterosexual mounting and a post-operative decline in isosexual mounting, the experimental males maintained high levels of isosexual mounting (both single and serial mounts) and low levels of heterosexual mounting throughout the post-operative period. Two factors apparently combined to produce the castrated males' isosexual eroticism: (1) the retention of a juvenile-type isosexual orientation after gonadectomy, and (2) the lack of stimulation by APR female groupmates. Among the control males, dominance rank and copulatory frequency were positively and significantly correlated across the entire post-operative period, but no correlation was found between copulatory frequency and testosterone levels.

Ovariectomy depressed the mounting behavior of the experimental females. They mounted groupmates much less often than did control females. An increase in female-female mounting among the controls during the last two years of the study may have been related to an increase in menstrual cycling and/or the amount of time females spent in estrus after conceptions had been prevented by vasectomizing the males.

VI. Penile Manipulations and Masturbation

Definitions and Methods

Autoerotic behavior occurred regularly within both groups, and was recorded during both ad libitum and scan sampling. Masturbation by females was difficult to identify with certainty, although a few obvious cases were seen in each group (usually involving a female sliding her genitals back and forth against a stone or another monkey [1]; see also CHEVALIER-SKOLNIKOFF's [29] description of genital stimulation during female-female mounts by M. arctoides). Male autoerotic behavior, in contrast, was easily recognized since it usually involved obvious manipulation of the penis. The present chapter will therefore be concerned exclusively with male autoerotic patterns. The following behaviors were recorded.

1. Penile manipulation: Any combination of manual, oral or pedal self-stimulation of the penis, whether or not the motor patterns of ejaculation resulted. Manual and pedal stimulation usually involved tugging or stroking the penis [25]. Simply holding the penis was not counted as manipulation. Occasionally, manual stimulation involved unusual patterns, such as rubbing the penis with a stone or other object. Oral stimulation involved mouthing or licking the penis. Penile manipulation was recorded regardless of the presence or absence of an erection.

2. Masturbation: This pattern was defined as penile manipulation culminating in the motor patterns of ejaculation.

Besides stimulating their own genitals, males were occasionally seen manipulating their groupmates' penises. Such interactions were always recorded and usually involved manual stroking or tugging and/or mouthing the partner's penis. Presence or absence of an erection was irrelevant for the recording of inter-male penile manipulation. Instances of one male grooming another's penis were not counted as manipulation. No instances of one female obviously stimulating another female's genitals were seen during the study.

Results: Male Self-stimulation

The two populations of males showed similar rates of autoerotic behavior pre-operatively. Between 26 January-21 March 1972, the controls and experimentals averaged 0.01 and 0.02 instances of penile manipulation/male/h, respectively. The difference was not tested for significance due to the small sample sizes.

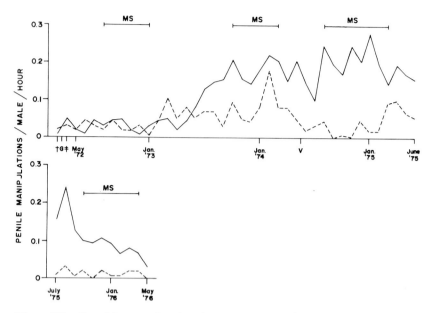

Fig. 29. Penile manipulation episodes/male/hour (combined data from scan and _ad libitum_ samples). Control males (all) _____, castrated males (all) - - - - - - - - -. † 26 January–21 March 1972, ‡ 27 March–April 1972, G gonadectomies, V vasectomies, MS mating season.

After the gonadectomies, penile manipulation occurred at similar rates within the two groups for just over a year (fig. 29). During the summer of 1973, however, as the monkeys were approaching mating season 2 (the first mating season during which all control males were post-pubertal), the intact males' penile manipulation rate rose sharply to a level considerably higher than that of the experimental males. As shown in figure 29, the controls then proceeded to engage in more penile manipulation than the castrates for the remainder of the study. Between 27 March 1972–May 1976, control males were observed to manipulate their own penises on 1160 occasions (mean rate of 0.11 manipulation episodes/male/h). In contrast, only 447 instances of self-stimulation were seen among the castrates during the same period, for a mean rate of 0.04 episodes/male/h (Mann-Whitney U-test, p < 0.0001). A total of 166 instances of masturbation were observed among the controls (14.3% of the penile manipulation total), for a mean post-operative masturbation rate of 0.016 episodes/male/h. One possible case of masturbation was observed within the experimental group. This instance involved male 241 and occurred 11 months after his castration.

In a few instances, males were seen using stones or other objects as part of penile manipulation. For example, control male 266 and castrated male 330 were seen to rub small stones against their penises on one occasion each. Male 330 was also involved in perhaps the most bizarre episode of penile manipulation of the study, during which he rubbed a small hermit crab against his erect penis. 330 repeatedly allowed the crab to attach itself (apparently with its pincers) to his penis, only to immediately detach it. When he abandoned the crab, after about 6 minutes of "stimulation," 330's penis appeared red and raw.

Spearman rank correlation coefficients were used to determine whether the control males' autoerotic behaviors were related to their copulatory frequencies. Non-significant correlations for each individual mating season and for the summed mating seasons indicated no relationship between the frequencies of penile manipulation and copulation. Similarly, Spearman coefficients suggested no relationship between autoerotic behavior and testosterone levels among the controls.

Results: Inter-male Penile Manipulation

Although the control males engaged in more frequent autoerotic behavior than the castrates, the experimental monkeys showed a much higher rate of inter-male penile manipulation. During pre-operative observations (26 January-21 March 1972), no instances of one male manipulating the penis of another were observed in either group. Across the post-operative period, however, five such instances were seen within the control group, while 35 instances were recorded for the castrated males (mean rates of 0.0001 and 0.001 inter-male penile manipulations/male-male dyad/h, respectively). A Chi-square test indicated that the difference between the rates was not due to chance ($\chi^2 = 21.9$, df 1, $p < 0.001$). This suggests that the castrates were significantly more likely to engage in this sort of behavior than the controls.

The five episodes of inter-male penile manipulation observed within the control group involved three different males as manipulators and four different males as recipients. All of the episodes featured only manual stimulation of the partner's penis, and none resulted in ejaculation. The 35 inter-male manipulation episodes recorded for the castrates featured seven different males as manipulators and eight males as recipients. In contrast to the control data, only 10 of the experimentals' inter-male episodes were limited to manual stimulation (fig. 30), with the remaining 25 cases (71.4%) including fellatio. As was true among the controls, none of the castrates' inter-male manipulation episodes resulted in ejaculation motor patterns.

An analysis of the relationship between dominance rank and the role (manipulator or recipient) assumed during inter-male penile manipulation found that all five of the controls' epi-

Fig. 30. Inter-male penile manipulation within the experimental group. Male 231 reaches for and tugs on male 301's penis.

sodes featured the dominant monkey as manipulator and the subordinate as recipient. A two-tailed Binomial Test suggested a link between the variables (p = 0.062). Such a relationship was confirmed when 67.6% of the castrates' inter-male manipulation episodes (23 of the 34 instances in which the participants' relative ranks were known) were found to feature the dominant male as the manipulator. These results were probably not due to chance (χ^2 = 4.24, df 1, p < 0.05).

Discussion

Penile manipulation and masturbation have been reported for numerous Old World monkey species (rhesus [25, 112, 129]; M. fuscata [71]; M. arctoides [18]; Cercopithecus aethiops [158]; Erythrocebus patas [J. LOY, unpub. obs.]). Although under some circumstances such behavior may be a substitute for copulation when the latter is impossible [112, 158], in many cases males have been observed to masturbate in preference to copulating with available receptive females (rhesus [25]; M. fuscata [71]; E. patas [J. LOY, unpub. obs.]). After their mid-1973 increase in penile manipulation frequency, the present study's control males maintained high levels of the behavior through the 1975 non-mating season, with little evidence of seasonal cycles (fig. 29). During mating season 4, however, the controls showed a sharp and inexplicable drop in penile manipulation frequency. These data (plus the non-significant Spearman correlation between copulatory frequency and penile manipulations for mating season 4) suggested that non-copulating control males (see chapter V) were not turning to masturbation for sexual satisfaction.

The low rate of penile manipulation shown by the castrated males may have been caused by a combination of factors including low testosterone levels, a lack of sexually stimulating females, a lack of reward (i.e. orgasm) during past penile manipulations (see MICHAEL and WILSON [114] for a similar argument with regard to the loss of mounting behavior among castrates), and the small size of their penises. With regard to the last variable, it appeared on several occasions that the castrated males had considerable difficulty extracting their small penises from the folds of the prepuce and scrotum. Several instances were observed in which a castrated male went to considerable effort to extract his flaccid penis only to cause it to disappear back into the preputial/scrotal folds by tugging on his scrotum as he lowered his head toward his groin. Usually the male would then set to work to extract his penis once again.

The rate of inter-male penile manipulation was 10 times higher among the castrates than among the controls. These results are of considerable interest because they support the theory that the castrated males developed unusually strong isosexual erotic attachments. The significance of the fact that the castrated males frequently stimulated one another orally while the control males did not is unclear. Both manual and oral stimulation of the penis (occasionally mutually given) have been observed among stumptail monkeys (M. arctoides [29, 30]). Interestingly, among stumptails male-male penile manipulations generally accompany "presenting" or mounting, while among the castrated males of the present study, such interactions generally took place between grooming episodes or while the partners were quietly sitting together.

Finally, the observation that dominants generally manipulated the penises of subordinates (and not vice versa) was probably related to the fact that dominant monkeys can inspect or manipulate the ventral surface of a subordinate with little fear of being attacked. Subordinates, on the other hand, might be expected to prefer contact with the dorsal, or "safe," surfaces of dominant animals. Thus, subordinates would rarely manipulate the penises of males who out-rank them.

Summary

The control males showed much higher post-operative rates of sexual self-stimulation (penile manipulation and masturbation) than the castrated males. No correlations were found between the control males' autoerotic activities and their copulatory frequencies or testosterone levels. Female autoerotic behavior probably occurred within both groups, but was difficult to identify with certainty, and therefore impossible to quantify accurately.

Instances of one male manipulating the penis of another were seen in both groups, but occurred significantly more often among the castrates. This finding was taken as additional evidence for the development of strong isosexual erotic attachments among the castrated males.

VII. Grooming

Definitions and Methods

A common characteristic of Old World monkeys is frequent grooming between groupmates. Such allogrooming interactions (as they have been labeled by SPARKS [157] and others) are thought to have important hygienic effects [80]; to reduce tension between animals [162], as in the case of placating actual or potential aggressors [146]; and to aid in the development and maintenance of social bonds [98, 146, 149, 151]. From the beginning of our study, grooming was one of the most frequent interaction patterns shown by the La Parguera monkeys [100]. Data on grooming were collected both pre- and post-operatively during ad libitum sampling, and post-operatively using the scan sampling technique.

Grooming (allogrooming)* was defined as the examination by an animal of the body/fur of a groupmate (using the slide, scrape, and pick movements described by SADE [146]) including the removal of parasites, scabs, dirt, etc. Episodes of grooming varied greatly in duration. Some lasted only a second or two and consisted of a single slide or scrape. Most, however, lasted for several seconds. An episode was considered terminated if (a) the partners separated, (b) the partners remained together but grooming movements stopped for 3-5 seconds, or (c) the partners changed roles (groomer to groomee). During analysis, all episodes were treated equally regardless of duration.

Eleven grooming indices were calculated monthly. They included:

1. Grooming episodes/dyad/h
2. Male-male grooming (episodes)/male-male dyad/h
3. Male-female grooming/heterosexual dyad/h
4. Grooming given/male/h: For this index, male-male and male-female grooming episodes were totaled. The total was divided first by the number of males in the group and second by the month's total scan sampling hours.
5. Grooming received/male/h: Male-male and female-male grooming episodes were totaled. The sum was divided first by the number of males in the group and second by total scan hours.

* Rhesus monkeys often groom themselves, but such auto-grooming behaviors were not analyzed for this report.

6. Female-female grooming/female-female dyad/h
7. Female-male grooming/heterosexual dyad/h
8. Grooming given/female/h: Calculated similar to index 4, only for females.
9. Grooming received/female/h: Calculated similar to index 5, only for females.
10. Isosexual grooming/isosexual dyad/h
11. Heterosexual grooming/heterosexual dyad/h

Within the control group, grooming episodes involving infants occurred between April 1973 and the spring of 1974. These interactions were tallied separately on the scan sample checksheets and were not analyzed for this report.

Results: Pre-operative Grooming

As noted by LOY and LOY [100], the patterns of grooming observed pre-operatively within the group of 33 juveniles were similar to those seen in free-ranging rhesus groups. Grooming was strongly influenced by kinship; relatives preferred each other over non-relatives as grooming partners. In addition, dominance influenced grooming. Mid-ranking and high-ranking monkeys were groomed more often than they groomed others. Sex and age also affected grooming; the 1969 monkeys tended to have more grooming partners than the 1970 animals, and females tended to have a greater number of partners than males. Finally, sexual precocity affected the grooming behavior of certain animals destined to become controls. For example, male 399 series-mounted female 242 on several occasions between 3 December 1971-21 January 1972, and during that same period 76 grooming episodes were recorded for the pair.

During the brief pre-operative period following the formation and separation of the control and experimental groups (26 January-21 March 1972), inter-group differences in grooming were small. The control males groomed all groupmates at the rate of 0.60 ± 0.15 episodes/male/h, which was not significantly different from the experimental males' mean rate of 0.69 ± 0.10 episodes/male/h (Mann-Whitney U-test, $p > 0.05$). Similarly, the inter-group difference in overall female grooming rates was non-significant (controls 1.03 ± 0.19 vs. experimentals 1.32 ± 0.21 episodes given/female/h; Mann-Whitney U-test, $p = 0.221$). Within both groups, females groomed other monkeys more frequently than did males. Heterosexual grooming rates were nearly equal in the two groups, as were the rates of male-male grooming. In contrast, the experimental females showed a higher rate of inter-female grooming (0.21 episodes/dyad/h) than the controls (0.13 episodes/dyad/h). A Chi-square test indicated that the inter-group difference in grooming between females was probably not due to chance ($\chi^2 = 35.83$, df 1, $p < 0.001$).

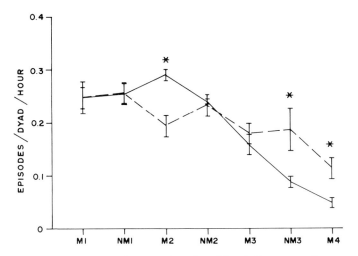

Fig. 31. Grooming episodes/dyad/hour. Mean ± SE. Controls (all) _____, experimentals (all) - - - - - - - -, significant inter-group difference shown by *.

Results: Post-operative Grooming

Grooming/dyad/h

Across the entire period of scan sampling (July 1972-May 1976), the two groups did not differ significantly in overall grooming rates (controls 0.18 ± 0.01 vs. experimentals 0.20 ± 0.01 episodes/dyad/h; $F = 0.680$, df 1/90, $p > 0.05$). A seasonal analysis revealed a general post-operative decline in grooming within both groups, as well as three significant seasonal differences (fig. 31). As will be shown, the controls' relatively high value during mating season 2 was due to a peak in heterosexual grooming, while the experimentals' significantly larger values during the last two seasons were due to frequent isosexual interactions. Despite these individual differences, the two groups did not differ significantly across the summed mating or summed non-mating seasons in grooming/dyad/h.

Male-male grooming/dyad/h

Although the pre-operative rates of male-male grooming were similar for the two groups, just a few months after their castrations the experimental males were grooming each other significantly more often than were the intact males. As shown in figure 32, the castrates exceeded the controls in intermale grooming during all seven post-operative seasons. Consequently, the experimental males' values for the total post-

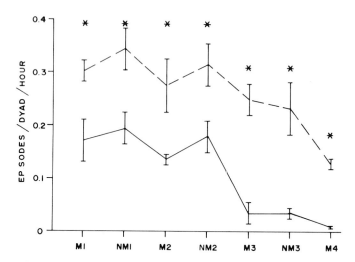

<u>Fig. 32.</u> Male-male grooming episodes/dyad/hour. Mean ±
SE. Control males (all) _____, castrated males (all)
- - - - - - -, significant inter-group difference shown by *.

operative period (castrates 0.26 ± 0.02 vs. controls 0.11 ±
0.01; F = 51.45, df 1/90, p < 0.001), for the summed mating
seasons (F = 34.443, df 1/50, p < 0.001), and for the summed
non-mating seasons (F = 20.185, df 1/38, p < 0.001) were significantly larger as well. Both male populations may have
shown weak seasonal cycles in male-male grooming, with higher
rates during the non-mating seasons. Neither group, however,
showed the expected rise during non-mating season 3. Inter-
male grooming declined in both groups as the monkeys grew
older.

<u>Male-female grooming/dyad/h</u>
Given the numerous other indications of the castrated
males' isosexuality, we were not surprised to discover that
they groomed each other frequently. We were somewhat surprised, however, to find that they also groomed females as frequently as did their control counterparts. Across the total
post-operative period, both controls and experimentals averaged
0.06 male-female grooming episodes/dyad/h (SE's 0.006 and
0.004, respectively; F = 0.512, df 1/90, p > 0.05). Inter-
group differences across the summed mating and across the summed non-mating seasons were non-significant. Furthermore,
there was only one significant seasonal difference (fig. 33).
The controls showed rather weak seasonal cycles in male-female
grooming, with higher values during the mating periods. They
did not, however, show a rise during mating season 4.

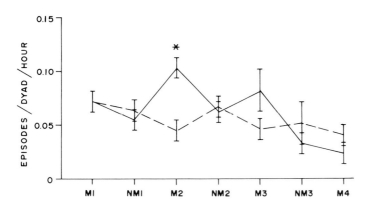

<u>Fig. 33.</u> Male-female grooming episodes/dyad/hour. Mean ±
SE. Controls (all) _____, experimentals (all) - - - - -
- - - - - - - -, significant inter-group difference shown by *.

<u>Grooming given/male/h</u>
As expected from the above data, this general index indicated that the castrates clearly surpassed the control males
in total grooming given to groupmates. The castrates' values
were significantly higher across the entire post-operative
period (1.30 ± 0.08 vs. 0.80 ± 0.07 episodes given/male/h; F =
21.451, df 1/90, p < 0.001), across the summed mating seasons
(F = 8.187, df 1/50, p < 0.01), and across the summed non-mating seasons (F = 14.153, df 1/38, p < 0.001). A seasonal analysis indicated significantly higher values for the experimentals during every season except mating periods 2 and 3, when
the controls peaked in male-to-female grooming (fig. 33).

<u>Grooming received/male/h</u>
In contrast to the strong inter-group differences in
grooming given by males, the data on episodes of grooming re-
ceived/male/h indicated similar levels for controls and castrates. A seasonal analysis revealed significant differences
only during mating seasons 2 (controls over castrates) and 4
(castrates over controls). Analyses across the entire post-
operative period, across the summed mating seasons, and across
the summed non-mating seasons showed non-significant inter-
group differences. As will be shown below, the similarity in
grooming-received by the two male populations was related to
the fact that high rates of isosexual grooming among the experimentals were balanced by more female-to-male grooming among
the controls.

<u>Female-female grooming/dyad/h</u>
Grooming between females occurred at roughly similar rates
in the two groups during much of the study. The experimental
females did show a higher overall rate (total post-operative

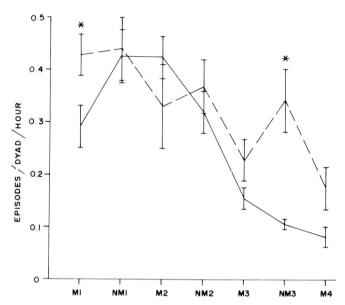

Fig. 34. Female-female grooming episodes/dyad/hour. Mean ± SE. Control females (all) _____, experimental females (all) - - - - - - - -, significant inter-group difference shown by *.

period: experimentals 0.32 ± 0.02 vs. controls 0.25 ± 0.02 episodes/dyad/h; F = 4.147, df 1/90, p < 0.05), due primarily to significant differences during mating season 1 and non-mating season 3 (fig. 34). Neither the summed mating nor the summed non-mating seasons produced significant inter-group differences in female-female grooming.

The experimental females (but not the controls) showed seasonal cycles in inter-female grooming, with higher levels during the non-mating seasons (fig. 34). Interestingly, this was the same sort of cycle shown by both male populations in inter-male grooming (fig. 32).

Female-male grooming/dyad/h

The experimental females groomed the castrated males at low, but uniform, rates across the post-operative period (fig. 35). In contrast, the intact females showed relatively high rates of grooming males during the first four post-operative seasons, followed by a strong decline in this behavior. The early differences were sufficient to carry the total post-operative and summed seasonal comparisons in the controls' favor (total post-operative: controls 0.13 ± 0.011 vs. experimentals 0.07 ± 0.004 episodes/dyad/h; F = 30.117, df 1/90, p < 0.001; summed mating seasons: F = 15.548, df 1/50, p < 0.001;

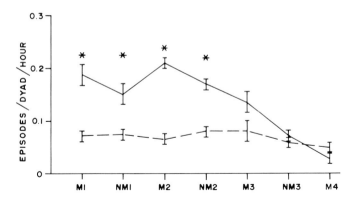

<u>Fig. 35.</u> Female-male grooming episodes/dyad/hour. Mean ±
SE. Controls (all) _____, experimentals (all) - - - - - -
- - - - - - - - -, significant inter-group difference shown by *.

summed non-mating seasons: F = 16.011, df 1/38, p < 0.001).
Within both groups, females groomed each other much more frequently than they groomed males (compare figures 34 and 35).
  The experimental females showed no evidence of seasonal
cycles in grooming males, while the control females tended to
show mating season peaks early in the post-operative period.
This pattern, however, was lost during mating seasons 3 and 4.

  Grooming given/female/h
  As the post-operative period progressed, all forms of female allogrooming declined within both groups. The control
females groomed others significantly more frequently than did
the experimental females during mating season 2 (F = 22.216,
df 1/10, p < 0.001) and during non-mating season 2 (F = 7.202,
df 1/6, p < 0.05). The first difference was related to coincidental highs in the controls' inter-female and female-male
grooming; the second was due solely to a peak in females grooming males. Despite these two seasonal differences, inter-
group comparisons across the total post-operative period,
across the summed mating seasons, and across the summed non-
mating seasons all produced non-significant results.

  Grooming received/female/h
  Due primarily to declines in female-female grooming (fig.
34), the number of grooming episodes received/female/h became
progressively smaller across the post-operative period within
both groups. The single significant seasonal difference occurred during mating season 2 and favored the controls (F =
5.237, df 1/10, p < 0.05). It was caused by coincidental highs
in heterosexual and isosexual grooming (fig. 33, 34). Inter-
group differences across the entire post-operative period,

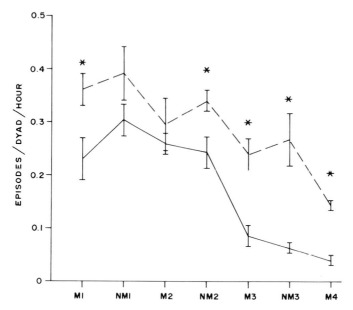

Fig. 36. Isosexual grooming episodes/dyad/hour. Mean ±
SE.   Controls (all) _____, experimentals (all) - - - - -
- - - - - - - -, significant inter-group difference shown by *.

across the summed mating seasons, and across the summed non-
mating seasons were all non-sigificant.
     The experimental females' tendency to groom each other
more frequently during the non-mating seasons than the mating
seasons (fig. 34) produced similar seasonal cycles with regard
to total grooming received/female/h.

Isosexual grooming/dyad/h
     As expected from the above data, the experimentals' over-
all rates of isosexual grooming were considerably higher than
those for the controls. Across the total post-operative period, the experimentals averaged 0.28 ± 0.02 grooming episodes/
isosexual dyad/h, compared to the controls' mean of 0.17 ±
0.02 (F = 20.677, df 1/90, p < 0.001). Significant differences were also found across the summed mating seasons (F =
12.921, df 1/50, p < 0.001), across the summed non-mating seasons (F = 8.874, df 1/38, p < 0.01), and in five of seven individual reproductive seasons (fig. 36). In each instance, the
experimental monkeys exceeded the controls.
     The weak seasonal cycles displayed in the experimentals'
male-male and female-female grooming interactions were quite
clear when total isosexual grooming was analyzed (fig. 36).
Seasonal cycles were not shown by the controls.

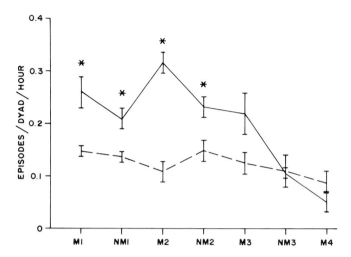

Fig. 37. Heterosexual grooming episodes/dyad/hour. Mean
± SE. Controls (all) _____, experimentals (all) - - - -
- - - - - - -, significant inter-group difference shown by *.

Heterosexual grooming/dyad/h
    Due primarily to frequent grooming of intact males by
their female groupmates, the controls exceeded the experimentals in overall heterosexual grooming across the total post-
operative period (0.19 ± 0.02 vs. 0.12 ± 0.01; F = 17.119, df
1/90, p < 0.001), across the summed mating seasons (F = 12.298,
df 1/50, p = 0.001), and across the summed non-mating seasons
(F = 4.897, df 1/38, p < 0.05).   Analyses of individual reproductive periods revealed four significant seasonal differences
in favor of the controls. All of them occurred during the
early and middle portions of the post-operative period (fig.
37). The cycles in heterosexual grooming (mating season highs
and non-mating season lows) shown early by the controls disappeared when their heterosexual grooming plummeted after
mating season 3.

Results: Grooming Behavior of Female 370

    In order to determine the effects of cycling female 370 on
the experimental monkeys' grooming behavior, portions of the
ad libitum grooming data were analyzed. Stratified random
samples of 100 grooming episodes involving 370 were drawn from
the summed mating seasons (25 episodes from each season) and
from the summed non-mating seasons (33 episodes each from non-
mating seasons 1 and 2, and 34 episodes from non-mating season
3). Similar samples were drawn for the other experimental fe-

Table XIII. Female 370's grooming behavior

Season	Subject(s)	Sample size	% Isosexual	% Heterosexual
Summed mating	370	100 episodes	21	79
	other experimental females	100 episodes	60	40
Summed non-mating	370	100 episodes	31	69
	other experimental females	100 episodes	69	31

males combined. Individual episodes were then categorized as isosexual, male-female, or female-male.

Analyses of these data revealed that regardless of season 370 was involved in a much higher percentage of heterosexual grooming episodes than her female groupmates. Furthermore, 370 showed somewhat more heterosexual grooming during the mating seasons than during non-mating periods (table XIII). Interestingly, there was little seasonal change in the percentages of grooming that 370 gave to and received from males. During the mating seasons, 57% of her heterosexual grooming was male-to-female and 43% was female-to-male. Corresponding figures for the non-mating seasons were similar, with 58% of her heterosexual grooming being male-female and 42% being female-male.

Discussion

The results of the grooming analyses indicated that the experimental monkeys (especially the males) were as isosexually oriented in this non-reproductive social behavior as they were with regard to mounting. Across the post-operative period, the experimentals showed over twice as many isosexual grooming episodes/dyad/h as heterosexual episodes (compare figures 36 and 37). In contrast, the controls showed slightly more heterosexual than isosexual grooming, although both forms were frequent.

Both the experimental males and females showed more isosexual grooming than their control counterparts. The intergroup difference was, however, greater and more consistent for males (compare figures 32 and 34). Many of the grooming episodes between castrated males may have been the aftereffects of agonistic interactions. It is not unusual for a defeated

rhesus monkey to approach and groom its aggressor shortly after
a fight [146]. Presumably such behavior aids in the repair of
strained relationships. Given the frequent fighting among the
castrated males (fig. 10), reconciliation via grooming may have
been a common occurrence throughout the study. It is clear,
however, that only a portion of the experimentals' male-male
grooming was related to agonism. The maintenance of relatively high levels of inter-male grooming during periods of infre-
quent isosexual fighting (e.g. mating season 3; compare figures
10 and 32) indicated that grooming occurred regularly between
same-sex "friends," and not simply as a response to strained
relationships. This last conclusion appeared to apply to females as well as males.

Due to their relatively high levels of isosexual grooming,
the experimental males showed significantly greater mean rates
of total grooming given (male and female recipients combined)
than the control males (recall that male-to-female grooming was
about the same in both groups). This finding is in contrast to
the report by BIELERT [19] that neo-natally castrated rhesus
males, raised and tested with intact females, groomed others
at about the same rate as did intact males under the same circumstances. Whether the conflicts between BIELERT's results
and those of the present study are due to differences in ages
at castration, group composition, or rearing/testing conditions
is unknown.

Within both groups, females groomed others more often than
did males across the total post-operative period (controls:
males 0.80 ± 0.07 vs. females 1.84 ± 0.16 episodes given/mon-
key/h; experimentals: males 1.30 ± 0.08 vs. females $1.51 \pm$
0.10). These data are in agreement with reports that among
free-ranging rhesus monkeys, females are more frequent groomers
than males both as juveniles [95] and as adults [46]. Similar
data for juveniles have been reported for Cercopithecus aethiops
[133] and Miopithecus talapoin [183]. In many primate species
adult females generally exceed adult males in grooming [121,
122].

The sharp decline in the controls' female-female grooming
that occurred after mating season 2 (fig. 34) was difficult to
explain, but it could have been related to the removal of all
infants in early 1974 (table IV). Rhesus females with infants
attract other females, who approach and groom the mother and/
or baby [95; J. LOY, pers. obs.]. (This phenomenon is common
to many monkey species, including baboons (Papio cynocephalus
[67]) and bonnet macaques (Macaca radiata [152]).) It is possible that removing the control infants eliminated much of the
stimulus for female-female grooming and/or disrupted inter-female relationships.

Another unexplained change in the controls' grooming was
the decline in female-male grooming that followed non-mating
season 2 (fig. 35). Although the control females were coming
into estrus and mating (see chapter V), and although estrous
females tend to groom males more frequently than do anestrous
females [95], peaks in female-male grooming failed to develop

Table XIV. Heterosexual grooming by and of control males during mating season 4*

Male	Male-female grooming		Female-male grooming		Total hetero-sexual episodes	% of grand total N
	N	episodes per h	N	episodes per h		
266	106	0.85	192	1.54	298	44.02
334	78	0.63	32	0.26	110	16.25
other control males (N = 6)	105	0.14**	164	0.22**	269	39.73
			Grand total N = 677		677	100.00

* Data taken from 124.75 h of ad libitum observations between October 1975-April 1976.
** episodes/male/h

during mating seasons 3 and 4. The grooming decline occurred
soon after the control males' vasectomies (performed on 30 May
1974). We have, however, no reason to think that the two events
were related.

Among the controls, both forms of heterosexual grooming
showed seasonal cycles early in the post-operative period (fig.
33, 35, 37). Male-female grooming showed peaks during mating
seasons 1-3, while female-male grooming peaked during the first
two mating periods. Increased heterosexual attraction as females passed through estrus was probably responsible for the
mating season peaks [110], which have also been reported to
occur among free-ranging rhesus monkeys [46, 86, 95, 98]. The
reason for the controls' failure to show a peak in male-female
grooming during mating season 4 was not clear, but could well
have been related to the domination of copulatory activity by
the alpha and beta males (see chapter V). By preventing the
other males from forming consort relationships (or "consocia-
tions" [182]), alpha male 266 and beta male 334 apparently precluded a typical mating season peak in heterosexual grooming
[95]. Together the top two control males accounted for just
over 60% of all heterosexual grooming during the final mating
season (table XIV).

Within both groups, male-male grooming showed weak seasonal cycles, with peak values occurring during the non-mating
seasons (fig. 32). For reasons unknown, both groups failed to
show a rise in male-male grooming during the final non-mating
season. Although most studies of free-ranging rhesus monkeys
have reported that adult males rarely groom each other, regardless of the season [46, 95], KAUFMANN [86] found a relative increase in such grooming partnerships during the non-mating
(birth) season on Cayo Santiago. The causes of non-mating season peaks in inter-male grooming are unclear. Perhaps an increase in friendly interactions between males naturally accom-
panies a non-mating season reduction in inter-male aggression.
This explanation does not appear to be applicable to the present study, however, since neither male population showed seasonal cycles in aggressiveness (compare figures 10 and 32).

Interestingly, the experimental females showed more obvious seasonal cycles in isosexual grooming than either male
population (fig. 34). Such cycles did not occur among the control females and have not been reported for free-ranging rhesus
monkeys [46]. Since the experimentals' peaks in inter-female
grooming were correlated with weak peaks in female-female
fighting (fig. 14), the "relationship repair" theory could be
used to explain grooming behavior. This is not particularly
satisfying, however, and on the whole the observed seasonal
cycles in female-female grooming remain unexplained.

In closing, a word should be said about the grooming behavior of experimental female 370. She participated in heterosexual grooming more frequently than the other experimental females, due no doubt to her hormonal and estrous cycles [95,
110]. If 370 had not been a member of the group, the experimentals' heterosexual grooming rates would certainly have been

somewhat lower. It should be kept in mind, however, that 370's grooming interactions with males were probably not due solely to sexual attraction. 370 was a high-ranking member of her group, and for that reason may have been a favorite grooming partner for both males and females. Based on a review of the literature, SPARKS [157] noted that most allogrooming among Old World primates goes from lower to higher-ranking individuals, and the model developed by SEYFARTH [151] suggests that this may be due to low-ranking animals attempting to establish/maintain supportive relationships with their superiors.

Summary

Allogrooming occurred frequently within both groups, although grooming rates tended to decline across the post-operative period. Most of the experimentals' grooming was isosexual, while much of the grooming by controls was heterosexual. As expected, the controls' heterosexual grooming tended to peak during the mating seasons (although this pattern was lost toward the end of the study). For reasons unknown, the experimentals' (and to some extent, the control males') isosexual grooming rates tended to peak during the non-mating seasons. Overall, the grooming data suggested that the isosexual orientation of the experimental monkeys (especially the castrated males) extended to non-reproductive social interactions as well as mounting.

VIII. Sitting-Touching

Definitions and Methods

Allogrooming is probably the interaction pattern most frequently used by investigators to identify and measure friendly relationships among monkeys. Nevertheless, several other types of interaction, including spending time in close proximity or in quiet body contact, are also thought to reflect positive inter-animal relations [8, 26, 147]. For this reason, instances of sitting-touching (defined below) were recorded throughout the present study during both ad libitum and scan sampling.

Sitting-touching was defined as dyadic body contact lasting at least three seconds and not involving grooming or other contact behaviors. Generally, the monkeys were relaxed and quiet during sitting-touching, and often leaned against one another. (It was common for animals to doze during sitting-touching episodes, especially during the heat of midday.) Extensive body contact was not required for an interaction to be counted; contacts as casual as crossed tails were recorded. Data on body contact involving the control infants were not analyzed for this report.

Fewer indices were needed to quantify sitting-touching than were used for other interaction patterns. This was because no records were kept of which animals initiated touching episodes, and body contact was therefore regarded as mutual, rather than unidirectional (e.g. male-to-female). Seven indices were calculated from the scan sample data:
1. Sitting-touching episodes/dyad/h
2. Male-male sitting-touching/male-male dyad/h
3. Female-female sitting-touching/female-female dyad/h
4. Isosexual sitting-touching/isosexual dyad/h
5. Heterosexual sitting-touching/heterosexual dyad/h
6. Sitting-touching episodes/male/h
7. Sitting-touching episodes/female/h

The calculations for these indices followed the formats already presented, with certain exceptions. Index 6 was calculated by adding the number of heterosexual episodes to twice the number of male-male interactions. That figure was divided first by the number of males in the group and second by the total scan hours. Index 7 was calculated similarly, only for females.

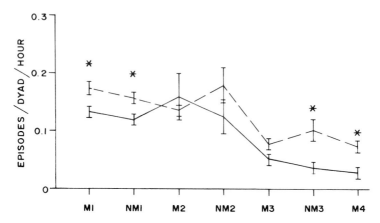

Fig. 38. Sitting-touching episodes/dyad/hour. Mean ± SE. Controls (all) _____, experimentals (all) - - - - - - - -, significant inter-group difference shown by *.

Results: Pre-operative Sitting-Touching

An analysis of the sitting-touching data from the pre-operative period immediately after group formation (26 January-21 March 1972) suggested the possibility of inter-group differences in social orientation prior to the gonadectomies. The experimental monkeys showed a mean overall rate of 0.07 ± 0.004 sitting-touching episodes/dyad/h, which was significantly greater than the controls' rate of 0.04 ± 0.005 (Mann-Whitney U-test, p < 0.004). This difference could have been related to the fact that the experimentals were in a smaller corral (0.09 ha, compared to 0.2 ha for the controls), and therefore had less space to spread out. In all likelihood, however, other variables also affected contact interactions. This fact is suggested by a division of the data into heterosexual and iso-sexual interactions. Although the two groups showed equal rates of heterosexual sitting-touching (means of 0.04 episodes/dyad/h), isosexual body contact occurred twice as frequently among the experimentals as among the controls (0.10 vs. 0.05 episodes/dyad/h; these means were taken from summed data and therefore SE's were not calculated nor was the difference tested for significance). Interestingly, the experimental females accounted for most of the isosexual sitting-touching within their group, showing 1.6 times as much same-sex contact as their male groupmates, and 3.25 times as much as the control females. The experimental males showed 1.6 times as much iso-sexual sitting-touching as their control counterparts.

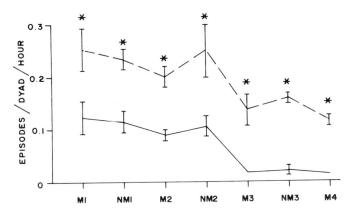

Fig. 39. Male-male sitting-touching episodes/dyad/hour.
Mean $\pm$ SE (missing SEs were less than 0.01). Control males
(all) _____, castrated males (all) - - - - - - - -, significant inter-group difference shown by *.

Results: Post-operative Sitting-Touching

Sitting-touching/dyad/h

The experimental monkeys' tendency to engage in more sitting-touching than the controls continued after the gonadectomies. The experimentals showed significantly more body contact
across the entire post-operative period (0.125 $\pm$ 0.01 vs. 0.091
$\pm$ 0.01 episodes/dyad/h; F = 8.125, df 1/90, p < 0.01) and
across the summed non-mating seasons (F = 8.576, df 1/38, p <
0.01), but not across the summed mating seasons. A seasonal
analysis indicated significant differences favoring the experimentals during the first two and last two seasons. No significant inter-group differences were observed during the middle of
the post-operative period (fig. 38).

Neither group showed clear seasonal cycles in sitting-
touching, and both groups showed some decline in body contact
over the post-operative period (fig. 38).

Male-male sitting-touching/dyad/h

As was true for grooming, the experimental males significantly exceeded the control males in inter-male body contact
during all seven post-operative seasons (fig. 39). The two
male populations differed significantly across the total post-
operative period (experimentals 0.187 $\pm$ 0.01 vs. controls 0.069
$\pm$ 0.01; F = 67.884, df 1/90, p < 0.001), across the summed mating seasons (F = 34.898, df 1/50, p < 0.001), and across the
summed non-mating seasons (F = 35.235, df 1/38, p < 0.001).

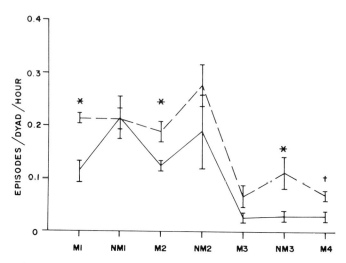

Fig. 40. Female-female sitting-touching episodes/dyad/
hour. Mean ± SE. Control females (all) _____, experimental females (all) - - - - - - - -, significant inter-group
difference shown by *. † Almost significant, p = 0.0503.

Both groups showed a post-operative decline in male-male
sitting-touching. By the end of the study it was a rare occurrence among the controls, although still relatively common
among the castrates (fig. 39). Neither group showed marked
seasonal cycles in male-male body contact.

Female-female sitting-touching/dyad/h
The experimental females sat-touching one another much
more frequently during the post-operative period than did the
control females (fig. 40). There were significant inter-group
differences favoring the experimentals during three individual
seasons, across the total period of scan sampling (0.157 ± 0.01
vs. 0.102 ± 0.01 episodes/dyad/h; F = 8.104, df 1/90, p < 0.01),
and across the summed mating seasons (F = 10.181, df 1/50, p <
0.01). Group means for the summed non-mating seasons were not
significantly different.
The control females' tendency toward increased isosexual
sitting-touching during the non-mating seasons disappeared after non-mating season 2, just as the experimental females were
developing weak cycles of this sort (fig. 40). In both groups,
the frequency of inter-female body contact declined sharply
during the final stages of the study.

Isosexual sitting-touching/dyad/h
Relatively high frequencies of both male-male and female-
female sitting-touching combined to give the experimentals much
higher levels of overall isosexual body contact than the con-

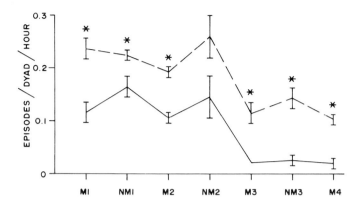

Fig. 41. Isosexual sitting-touching episodes/dyad/hour.
Mean ± SE (missing SEs were less than 0.01). Controls (all)
_____, experimentals (all) - - - - - - - -, significant
inter-group difference shown by *.

trols. Inter-group differences were significant during six of
seven post-operative seasons (fig. 41), as well as across the
entire post-operative period (experimentals 0.176 ± 0.01 vs.
controls 0.084 ± 0.01 episodes/dyad/h; F = 38.300, df 1/90, p <
0.001), across the summed mating seasons (F = 30.975, df 1/50,
p < 0.001), and across the summed non-mating seasons (F = 13.
306, df 1/38, p < 0.001).

Heterosexual sitting-touching/dyad/h

Although in many ways the control monkeys were clearly
more heterosexually oriented than the experimentals (e.g. series-mounting, grooming), this was not true with regard to sitting-touching. Analyses revealed non-significant inter-group
differences during all seven post-operative seasons (fig. 42),
as well as across the total post-operative period (experimentals 0.081 ± 0.01 vs. controls 0.096 ± 0.01; F = 1.204, df 1/
90, p > 0.05), across the summed mating seasons, and across the
summed non-mating seasons. Among the experimentals, heterosexual body contact showed no seasonal fluctuations and always
occurred less frequently than isosexual contact (compare figures 41 and 42). In contrast, the controls may have shown an
early tendency for mating season peaks and non-mating season
declines in heterosexual contact. This pattern was lost, however, after the second non-mating season. Among the controls,
heterosexual sitting-touching occurred somewhat more frequently
than isosexual contact during the mating seasons (although
there was little difference during mating season 4), but fell
to or below isosexual levels during non-mating periods (fig.
41, 42).

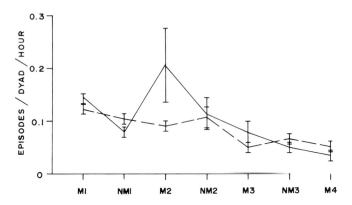

Fig. 42. Heterosexual sitting-touching episodes/dyad/hour.
Mean ± SE. Controls (all) _____, experimentals (all)
- - - - - - - -.

Sitting-touching episodes/male/h
    Due to their isosexual contacts, the castrated males
showed a much higher overall level of sitting-touching than did
the control males (total post-operative period: experimentals
1.913 ± 0.12 vs. controls 1.189 ± 0.13 episodes/male/h; F = 16.
489, df 1/90, p < 0.001). This difference held true across
both the summed mating seasons (F = 4.392, df 1/50, p < 0.05)
and the summed non-mating seasons (F = 18.016, df 1/38, p < 0.
001). Four out of seven seasonal comparisons revealed significantly greater values for the castrates than for the control
males (fig. 43).

Sitting-touching episodes/female/h
    Somewhat surprisingly, given their clear edge in isosex-
ual body contact, the experimental females significantly exceeded the control females in overall sitting-touching during
only one reproductive season (fig. 44). There were non-significant differences between the groups across the total post-
operative period (experimentals 1.637 ± 0.13 vs. controls 1.413
± 0.15; F = 1.331, df 1/90, p > 0.05), across the summed mating
seasons, and across the summed non-mating seasons.

Sitting-Touching Behavior of Female 370

    In order to determine whether or not cycling female 370
showed the same sort of contact interactions as the other experimental females, portions of the ad libitum data on sitting-
touching were analyzed. Stratified random samples of 100 sitting-touching episodes were drawn for both 370 and for the
other experimental females combined from (a) the summed mating

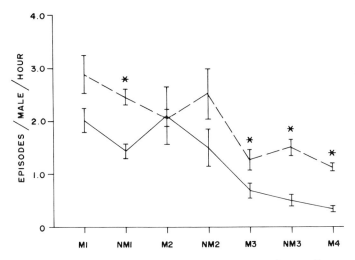

Fig. 43. Sitting-touching episodes/male/hour. Mean ± SE.
Control males (all) _____, castrated males (all) - - - -
- - - - - - - -, significant inter-group difference shown by *.

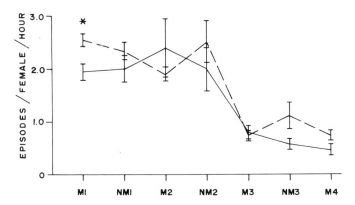

Fig. 44. Sitting-touching episodes/female/hour. Mean ±
SE. Control females (all) _____, experimental females
(all) - - - - - - - -, significant inter-group difference shown
by *.

Table XV. Sitting-touching interactions of female 370

Season	Subject(s)	Female-female contact	Heterosexual contact	Total episodes
Mating seasons	370	39*	61	100
	other experimental females	41	59	100
Non-mating seasons	370	39	61	100
	other experimental females	53	47	100

* Episodes of sitting-touching

seasons and (b) the summed non-mating seasons. Each interaction was categorized as isosexual or heterosexual, and the percentages for these two categories were calculated for each sample.

It was found that during the mating seasons, 370's ratio of isosexual/heterosexual body contact was very similar to that shown by her female groupmates (table XV). Furthermore, while 370's pattern of sitting-touching did not change from mating to non-mating seasons, the other experimental females apparently showed an increase in their percentage of isosexual contact during non-mating periods. In all, these results suggested that there were minor differences between 370's sitting-touching interactions and those of her female groupmates.

Discussion

The sitting-touching data provided additional support for the general conclusion that the experimental monkeys were more isosexually oriented than the controls. Both sexes of experimental animals showed higher frequencies of isosexual body contact across the total post-operative period (ratio of isosexual to heterosexual sitting-touching episodes/dyad/h: castrated males 2.31, experimental females 1.94). Among the controls, the females were only slightly more isosexual (isosexual/heterosexual ratio: 1.06) and the males were actually more heterosexually oriented (I/H ratio: 0.72).

The orientations reflected by the experimentals' contact interactions were similar (more so for males than for females) to data on free-ranging rhesus monkeys. SADE [147] studied seven Cayo Santiago animals (four males and three females) throughout their social development. When the monkeys were all two-year-olds, SADE found that the sitting-touching behavior of his three females varied from more female than male partners, to equal contact with the sexes, to more male partners. In contrast, all four juvenile males showed much more frequent body contact with males than with females (i.e. females other than their mothers). Therefore, with regard to the present study, it appears safe to conclude that the castrated males retained a juvenile pattern of body contact associations well into chronological adulthood, although the reasons for that retention are not completely clear.

Among the controls, some post-pubertal cyclicity in heterosexual sitting-touching was apparent early in the study, but disappeared after non-mating season 2 (fig. 42). The few observed cycles featured peaks in heterosexual contact during the mating seasons, when, due to females passing through estrus and males experiencing high testosterone levels, male-female attraction was at a maximum level [87, 94, 98] (also see CZAJA and BIELERT [40] for data on fluctuations in male-female proximity during the rhesus menstrual cycle). Precisely why the controls failed to show an increase in heterosexual body contact during mating season 3 is unclear, but it apparently was due primarily to a decline in female interest in males. This conclusion is based on the finding that during the third mating season there was a peak in male-to-female grooming among the controls, but no corresponding peak in female-to-male grooming (fig. 33, 35). The controls' lack of a peak in heterosexual sitting-touching during mating season 4 could have been related to the fact that the alpha and beta males monopolized mating behavior during that period, and apparently simultaneously suppressed several other forms of heterosexual interaction.

The cycle in female-female body contact shown early in the study by the controls (with non-mating season peaks and mating season declines) (fig. 40) could have been related to their pregnancies and/or the presence of infants in the group. Heavily pregnant rhesus females have been reported to be somewhat lethargic [94], and therefore may be especially prone to engage in sedentary activities such as sitting-touching. In addition, rhesus females are attracted to one another's infants. During the birth season (non-mating season) females with young often form "nursery groups" and participate in allogrooming and body contact [94, 95]. While these factors could have contributed to the controls' peak in female-female sitting-touching during non-mating season 1, the "nursery group effect" probably was not very important during the second non-mating period since the 1974 infants were removed from the control group shortly after birth. There was no increase in female-female body contact during the controls' third non-mating season (fig. 40), and, perhaps significantly, there were no pregnant females or new infants in the group at that time.

During the mid-to-late post-operative period, the experimental females developed the same sort of seasonal cycles in female-female sitting-touching that had characterized the controls (fig. 40). The basis for the development of a body contact cycle among these non-reproducing females is unknown.

Summary

Both the experimental males and females showed significantly more isosexual sitting-touching than their control counterparts across the post-operative period. The sustained high level of isosexual body contact among the castrated males appeared to reflect the retention of juvenile interaction patterns. Interestingly, the experimental monkeys showed about as much heterosexual sitting-touching as the controls, although the intact animals showed a weak seasonal cycle in heterosexual body contact while the experimentals did not. For reasons which are unclear, seasonal cycles in isosexual body contact were shown by the females of both groups.

IX. Play

Definitions and Methods

The young of virtually all primate species engage in frequent play [2, 44, 96, 121, 160], and this was certainly true
of our juvenile experimental and control monkeys. Although our
findings on play have been reported in previous publications
[100, 101], a summary will be presented here for several reasons. First, a review of the play data may help in understanding other behavior patterns. Second, the play data provide additional information about the monkeys' social orientations.
Finally, play provides perhaps the best proof that the experimental males continued to behave like juveniles into chronological adulthood.

Nonhuman primate play has the interesting property of being difficult to define precisely [154], but relatively easy to
recognize. Interactions classified as play during the present
study generally consisted of play-fighting behaviors [160] such
as chasing, wrestling, and mouthing. Vigorous play-fighting
was differentiated from actual fighting by the frequent rever-
sal of attacker and target roles during play and by the lack of
submissive facial expressions (grimaces) and agonistic vocalizations. Play interactions were generally dyadic; episodes
which involved several monkeys were reduced to dyadic combina-
tions for analysis. In order for an episode to be scored as
play, both partners had to be clearly playing. Play invitations, or unilateral attempts at play, which were not met with
a playful response from the target animal were not recorded as
play episodes. Similarly, no records were kept of solitary
play. The present chapter reports on social play only.

Play was recorded throughout the study during both ad
libitum and scan sampling. Seven play indices were calculated
from the scan sample data.
1. Play episodes/dyad/h
2. Male-male play episodes/male-male dyad/h
3. Female-female play episodes/female-female dyad/h
4. Isosexual play episodes/isosexual dyad/h
5. Heterosexual play episodes/heterosexual dyad/h
6. Play episodes/male/h
7. Play episodes/female/h
Detailed information on the calculation of the above indices was presented in an earlier paper [101]. It is sufficient here to note that the indices were calculated in a similar manner to those for grooming and sitting-touching, with one
significant exception. Play between control infants and adults

was <u>utilized</u> for indices 6 and 7. Infant-adult play was <u>not</u> used for the controls' indices 1-5.

With this introduction, a summary of the play behaviors seen during the study will be presented. Additional details, and charts illustrating the findings, can be found in our ear- lier report [101].

Results: Pre-operative Play

Observations made on the group of 33 juveniles (6 December 1971-21 January 1972) indicated very little difference between the play of the designated control monkeys and that of animals destined to be experimentals. There were non-significant dif- ferences in play/monkey/h between "control" and "experimental" males ($F = 0.375$, df 1/15, $p > 0.05$) and between "control" and "experimental" females ($F = 0.00$, df 1/14, $p > 0.05$). A clear inter-sex difference in overall play frequency was found, how- ever. Females played less than half as often as males.

During the period following group formation but prior to the gonadectomies (26 January-21 March 1972), some inter-group differences in play became apparent. Although the overall rates of play/dyad/h were about the same in the two groups ($F = 1.800$, df 1/4, $p > 0.05$), the experimental males were found to play significantly more than the control males (play/ male/h: 1.713 ± 0.11 vs. 1.203 ± 0.09; $F = 13.078$, df 1/4, $p < 0.05$). In contrast, the control females played more than their experimental counterparts (play/female/h: 0.380 ± 0.05 vs. 0.200 ± 0.03; $F = 11.435$, df 1/4, $p < 0.05$). The experimentals showed more frequent male-male play than the controls ($F = 19.00$, df 1/4, $p < 0.05$), but levels of inter-female play did not differ significantly between the groups. Perhaps due to the sexual precocity (and resultant heterosexual interest) of cer- tain of the control monkeys, their group rate of heterosexual play was twice that of the experimentals ($F = 8.000$, df 1/4, $p < 0.05$).

Results: Post-operative Play

Play/dyad/h
Within both groups, play occurred most frequently early in the study and declined as the post-operative period progressed. The decrease in play was, however, much greater among the con- trols than the experimentals. Across the post-operative peri- od the two groups differed significantly in total play (play/ dyad/h: experimentals 0.125 ± 0.01 vs. controls 0.061 ± 0.01; $F = 12.111$, df 1/90, $p < 0.001$). In addition, the experimen- tals played significantly more than the controls across the summed mating seasons ($F = 7.534$, df 1/50, $p < 0.01$) and across the summed non-mating seasons ($F = 4.388$, df 1/38, $p < 0.05$). There did not appear to be strong seasonal differences in play within either group.

Male-male play/dyad/h

The experimentals' pre-operative tendency to engage in
more male-male play than the controls continued after the
gonadectomies. Although inter-male play declined post-operatively within both groups, the drop was greater among the controls. The control males averaged less than one isosexual
play episode/h during the last year of the study, compared to
over six episodes/h among the castrates. The castrated males
significantly exceeded the controls in inter-male across the
total post-operative period (0.405 ± 0.04 vs. 0.204 ± 0.04
episodes/dyad/h; $F = 10.606$, df 1/90, $p < 0.01$) and across the
summed mating seasons ($F = 6.640$, df 1/50, $p < 0.05$). There
was no significant difference across the summed non-mating
seasons. Neither group showed seasonal cycles in male-male
play.

Female-female play/dyad/h

The ovariectomies appeared to have little effect on the
frequency of inter-female play. Across the total post-operative period, the experimental females averaged 0.026 ± 0.01
female-female play episodes/dyad/h, compared to the control females' rate of 0.017 ± 0.01 ($F = 1.548$, df 1/90, $p > 0.05$).
Inter-group differences across the summed mating seasons and
across the summed non-mating seasons were similarly non-significant. Neither group showed regular seasonal cycles in female-female play. Within both groups, inter-female play declined sharply during the first half of the post-operative
period, and rarely occurred during the last two years of the
study.

Isosexual play/dyad/h

Due primarily to frequent play interactions between castrated males, the experimental monkeys showed significantly
higher overall levels of isosexual play than the controls
across the total post-operative period (0.237 ± 0.02 vs. 0.115
± 0.02 episodes/dyad/h; $F = 13.266$, df 1/90, $p < 0.001$), across
the summed mating seasons ($F = 8.449$, df 1/50, $p < 0.01$), and
across the summed non-mating seasons ($F = 4.662$, df 1/38, $p <
0.05$). Because neither group displayed seasonal cycles in
inter-male or inter-female play, the same held true for the
general measure of total isosexual play.

Heterosexual play/dyad/h

The pre-operative edge shown by the controls over the experimentals in heterosexual play was quickly lost post-operatively. Due to strong declines in playfulness among both males
and females, heterosexual play among the controls plummeted
during the first two post-operative years. It was almost non-
existent during the final two years of the study. In contrast,
the experimentals showed a slower decline in heterosexual play
(due primarily to the continuing playfulness of the castrated
males) and significantly exceeded the controls during post-operative years 2 and 3. Nevertheless, the two groups did not

differ significantly across the entire post-operative period
(experimentals 0.026 ± 0.005 vs. controls 0.015 ± 0.004 epi-
sodes/dyad/h; F = 3.079, df 1/90, p > 0.05), across the summed
mating seasons, or across the summed non-mating seasons. Neither group showed marked seasonal differences in the rate of
heterosexual play.

Play/male/h

As expected from the above results, the castrated males
showed much higher overall levels of play than their control
counterparts across the entire post-operative period (3.110 ±
0.36 vs. 1.585 ± 0.33 episodes/male/h; F = 9.882, df 1/90, p <
0.01). Interestingly, although the castrated males significantly exceeded the controls across the summed mating seasons
(F = 6.272, df 1/50, p < 0.05), they did not do so across the
summed non-mating periods. By the last year of the study, the
castrated males still showed relatively high rates of play
(about eight times greater than the control males' rate), even
though they were chronologically adult (5.5-6.6 years of age).

Play/female/h

Both populations of females showed strong age-related
drops in play frequency, with play becoming rare by the last
two years of the study. There were non-significant inter-group
differences in play/female/h across the total post-operative
period, across the summed mating seasons, and across the summed
non-mating seasons.

Play Behavior of Female 370

There seemed to be little reason to suspect that the presence of cycling female 370 had a significant influence on play
among the experimental monkeys. Like all other females, 370
played less than like-aged males and showed a sharp reduction
in play as she matured. Because of 370's influence on other
heterosexual interactions (e.g. mounting and grooming), however, a check was made to see if she had contributed dispropor-
tionately to the experimentals' heterosexual play during post-
operative years 2 and 3.

Heterosexual play among the experimental monkeys was still
relatively frequent during post-operative year 2. Therefore, a
stratified random sample of 100 heterosexual episodes was drawn
from ad libitum data from that time period. In contrast, a
total of only 43 episodes of heterosexual play was recorded
during post-operative year 3, and so all episodes from that
period were analyzed. Chi-square test results suggested that
370 had no more tendency to engage in heterosexual play than
her acyclical female groupmates (year 2: χ^2 = 0.57, df 1, p >
0.30; year 3: χ^2 = 2.62, df 1, p > 0.10).

Discussion

All studies of rhesus monkey play are in agreement that the behavior is shown most commonly by immature animals and that as the monkeys approach and pass puberty, play frequencies decline sharply. For both males and females, play frequency appears to peak late in the first year or during the second year of life [78, 100, 127, 145, 160]. Although adults of both sexes do play, such interactions are fairly rare. They are generally directed toward a juvenile or infant, and are most characteristic of young adults [23]. In addition to the age-related decline in play, it is generally agreed that immature males tend to play more than immature females [127, 145, 160] (this is also true for Miopithecus talapoin [183], Cercopithecus aethiops [133], and Papio hamadryas [92]).

Several of these features characterized the play behavior of our control and experimental monkeys [100, 101]. Within both groups, males engaged in much more play than like-aged females. In addition, the control males and females and the experimental females all showed typical age-related drops in play frequency and rarely played after they were fully grown. In contrast, although they experienced a strong drop from the very high frequencies they showed as immatures, the castrated males continued to play at relatively high rates as adults. During the final post-operative year, the castrated males, who were then 5.5-6.6 years of age, engaged in 1.55 play episodes/male/h--a rate characteristic of the control males when they were 3.4-4.4 years old.

In addition to showing unusually high levels of play as adults, the castrated males continued to engage primarily in isosexual play. The fact that immature rhesus males tend to play mostly with other males has been reported by several investigators [73, 145, 160]. Thus, the play behavior of the adult castrates was similar to that of normal juvenile males in two ways: play occurred relatively frequently, and it tended to be isosexual.

Our results suggest that declining playfulness in rhesus females is almost solely a function of age. Gonadal hormone production and reproductive history seemed to have little effect on our females' play behavior. Among rhesus males, on the other hand, playfulness is probably lost initially as a function of age. The final decrements in peer play appear to coincide with (1) the attainment of adult levels of testosterone and (2) reproduction-related changes in inter-male relationships (i.e. intensified competition for dominance rank and mating opportunities, with a concomitant reduction in friendly interactions). The relative contributions of hormonal and social factors to the loss of male play are unclear, and the problem is complicated by the probability that many of the relevant social factors are themselves testosterone-dependent to some extent. Nonetheless, at present it seems unlikely that testosterone has a direct effect on male playfulness. The finding by GORDON, ROSE, and BERNSTEIN [59] that the adult males of a captive rhesus group showed an inverse relationship between testosterone and play (suggesting a direct hormonal effect) is countered by the fact that the present study's control males showed clear seasonal cycles in testosterone but no cyclicity in play. Furthermore, BIELERT [19] found that injecting two-year-old rhesus male castrates with testosterone propionate resulted in a significant _increase_ in their rate of play. In sum, although a direct effect of testosterone on male play cannot be ruled out completely, it seems more likely that the hormone broadly affects male-male relationships by producing increased tension between males and that this, in turn, leads to a reduction in many forms of friendly interaction, including play. A reasonable hypothesis with regard to our castrated males is that they continued to play with one another as adults because they had not experienced the normal post-pubertal attenuation of male-male social relations.

Summary

Play occurred frequently within both groups at the beginning of the study, but decreased in frequency as the monkeys grew older. Both the control and experimental females played less than their male groupmates. When the females did play, they did so about as often with males as with other females. The decline in female play appeared to be almost solely age-related; ovariectomizing the experimental females had little effect on their play behavior. Within both groups, males engaged primarily in isosexual play. Male-male play among the controls dropped to very low levels during the last half of the post-operative period. In contrast, although their isosexual play declined as they grew older, the adult castrated males continued to play with one another at rates characteristic of intact juvenile or sub-adult monkeys.

X. A Summary of Results and a Discussion of the Juvenile-to-Adult Transition

Introduction

The preceding chapters have provided a piecemeal description of the behaviors observed during our study. While this presentation was necessitated by the quantity and complexity of our results, a synthesis is needed in order to gain an overview of the behavioral effects of gonadectomy. In addition, our notion that the adult experimental monkeys continued to interact like juveniles needs to be evaluated against the background of broadly based data concerning the normal behavior of immature rhesus monkeys. To these tasks, we now turn.

The Behavioral Effects of Gonadectomy

Males

Statistical comparisons between the behavior of the control males and the castrated males are summarized in table XVI. Included in that table are the results of inter-group tests across the total post-operative period (generally the same as the period of scan sampling, July 1972-May 1976), across the summed mating seasons, and across the summed non-mating seasons.

As can be seen from the table, the castrated males were more aggressive than their control counterparts toward both male and female opponents. Further, the castrated males mounted each other, both singly and serially, more frequently than did the control males. The control males, on the other hand, showed a significantly higher rate of male-female series-mounting than the castrates, and accounted for all of the post-operative copulations. Although the controls showed more autoerotic penile manipulation than the castrates, the experimental males engaged in inter-male penile manipulation more often than the controls and accounted for all instances of isosexual fellatio.

With regard to interactions which were neither agonistic nor sexual, the castrated males showed significantly more grooming-given, sitting-touching, and play than the control males. Isosexual interactions accounted for the experimental males' higher scores in all three categories. No significant differences were found between the groups in the rates of heterosexual "social" interactions.

Table XVI. Behavioral summary for males

Variable	Group showing significantly higher value for:		
	Total post-operative period	Summed mating seasons	Summed non-mating seasons
A. Agonism			
1. Male-male fights/d/h*	Castrates	Castrates	Castrates
2. Male-female fights/d/h	Castrates	ns**	Castrates
3. Fights/male/h	Castrates	Castrates	Castrates
4. Victories/ subordinate/h	Castrates	xxx***	xxx
B. Mounting			
1. Male-male mounts/d/h	Castrates	Castrates	Castrates
2. Male-female mounts/d/h	ns	ns	ns
3. Mounts/male/h	Castrates	ns	Castrates
4. Mounts re-ceived/male/h	Castrates	Castrates	Castrates
C. Series-mount-ing			
1. Male-female series/d/h	Controls	xxx	xxx
2. Male-male series/d/h	Castrates	xxx	xxx
D. Penile Manipulations			
1. Self-manipu-lations/male/ h	Controls	xxx	xxx
2. Male-male manipulations/ d/h	Castrates	xxx	xxx

*   Male-male fights/dyad/hour
**   No significant inter-group difference
*** Inter-group difference not tested

Table XVI. (continued)

Variable	Group showing significantly higher value for:		
	Total post-operative period	Summed mating seasons	Summed non-mating seasons
E. Grooming			
1. Male-male episodes/d/h	Castrates	Castrates	Castrates
2. Male-female episodes/d/h	ns	ns	ns
3. Grooming given/male/h	Castrates	Castrates	Castrates
4. Grooming received/male/h	ns	ns	ns
F. Sitting-touching			
1. Male-male episodes/d/h	Castrates	Castrates	Castrates
2. Heterosexual episodes/d/h	ns	ns	ns
3. Episodes/male/h	Castrates	Castrates	Castrates
G. Play			
1. Male-male play/d/h	Castrates	Castrates	ns
2. Heterosexual play/d/h	ns	ns	ns
3. Play/male/h	Castrates	Castrates	ns

In addition to these inter-group differences, it is important to note that the control males showed clear seasonal cycles in male-female mounting, and less distinct seasonal fluctuations in heterosexual grooming and sitting-touching. Peaks in heterosexual interactions among the controls were generally concomitant with peak levels of circulating testosterone for the males and strong estrous periods for the females. Curiously, the low-hormone experimental monkeys showed weak seasonal cycles in male-female series-mounting which paralleled changes within the control group.

In sum, the castrated males can be described as having been more aggressive, more frequent mounters, and more socially active than the intact males. In addition, throughout the study, in all of their various interactions, the castrated males showed more of an isosexual orientation than the male controls.

The concepts of isosexual and heterosexual orientation were introduced in earlier chapters, but were not described in detail. Since these theoretical constructs are of central importance for our interpretation of the effects of gonadectomy, it is appropriate to digress briefly for a discussion of "orientation" before continuing with the account of the study's results.

Isosexual orientation can be defined as an attraction to, and preference for, same-sex partners. Conversely, heterosexual orientation is an attraction to, and preference for, opposite-sex partners. Obviously, orientation is a relative measure that is derived from observations of an animals' interactions. Within each major orientation type, at least three sub-types defined by particular interaction patterns may be differentiated:

1. Agonistic orientation: This sub-type pertains to aggressive/submissive interactions.

2. Erotic orientation: This category refers to the animal's sexual interactions (especially mounting).

3. Social orientation: This sub-type is related to such "friendly" interactions as grooming, sitting-touching, and play.

The concept of agonistic orientation may require some modification of the general definitions since animals are not usually thought of as being attracted to, or preferring, certain classes of opponents. Agonistic orientation is perhaps better defined as a tendency to be more attentative toward, and prone to interact aggressively with, members of one sex or the other. Nonetheless, it may not be entirely incorrect to define agonistic orientation in terms of partner preference. While animals probably prefer sexual and social partners of a particular sex because of the rewards derived from those interactions, they might equally as well prefer certain types of opponents because of the rewards derived from defeating key competitors.

Obviously, isosexual orientation and heterosexual orientation are not absolute, mutually exclusive, conditions. With regard to any interaction category (agonistic, erotic, social), animals usually are relatively more isosexually than heterosexually oriented (or vice versa), but rarely do they cease all interactions with one sex and devote themselves exclusively to the other. Orientation patterns, therefore, are a matter of degree. Furthermore, they may vary dramatically from one type of behavior to another. For example, as we will see, during the mating season it was common for the control males simultaneously to show a heterosexual erotic orientation and an iso-

sexual agonistic orientation. Finally, orientation patterns
often show temporal changes, sometimes becoming more intensely
isosexual or heterosexual than before, sometimes shifting per-
manently from one orientation to the other, and sometimes developing seasonal cycles of alternating isosexuality and heterosexuality.

Agonistic, erotic, and social orientation patterns for the
males of both groups were quantified across the study using
scan sample data (ad libitum data were used for the period from
26 January-21 March 1972). The general formula for the calculation of orientation score was:

$$\frac{\text{isosexual interaction rate} - \text{heterosexual interaction rate}}{\text{isosexual interaction rate} + \text{heterosexual interaction rate}}$$

This formula produced a range of values from +1.0 (complete
isosexuality) to -1.0 (complete heterosexuality). A score of
zero represented orientation equality. The interaction indices
used for calculating the males' orientation patterns were as
follows:

1. Agonistic orientation: Male-male fights/dyad/h and heterosexual fights/dyad/h.

2. Erotic orientation: Male-male mounts/dyad/h and male-
female mounts/dyad/h.

3. Social orientation (a summation of grooming, sitting-
touching, and play): Male-male grooming, sitting-touching, and
play/dyad/h and male-female grooming, heterosexual sitting-
touching, and heterosexual play/dyad/h.

Two examples will demonstrate how orientation scores were
calculated. During the first mating season, the castrated
males participated in 0.348 male-male fights/dyad/h and 0.172
heterosexual fights/dyad/h. Utilizing the formula given above,
an orientation score of 0.338 was obtained, a value in the middle of the isosexual range. In contrast, during mating season
3, the control males showed isosexual grooming, sitting-touching, and play rates of 0.036, 0.019, and 0.023 interactions/
dyad/h (sum = 0.078). The males' interaction rates for male-
female grooming, heterosexual sitting-touching, and heterosexual play were 0.083, 0.079, and 0.0, respectively (sum = 0.162).
Applying the formula produced a value of -0.35, which indicated
a moderate heterosexual orientation.

The orientation patterns shown by the control and experimental males are presented in figure 45. Both male populations
showed primarily isosexual agonistic orientations throughout
the study, although the castrated males' isosexuality was generally stronger than that of the controls. Neither group of
males showed seasonal cycles in agonistic orientation. With
regard to erotic orientation, the experimental males were
strongly isosexual across the entire study, while the post-pubertal control males showed seasonal cycles between relative
isosexuality (during the non-mating seasons) and relative heterosexuality (during the mating seasons). The failure of the

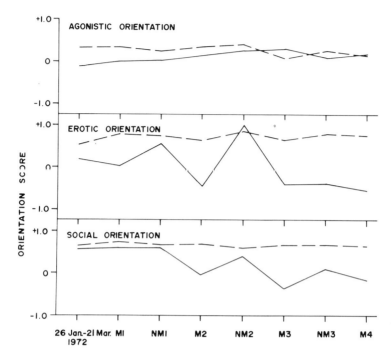

<u>Fig. 45.</u> Orientation patterns for males. Control males
(all) _____, castrated males (all) - - - - - - - -.
Orientation scores range from +1.0 (complete isosexuality) to
-1.0 (complete heterosexuality). See the text for an explanation of the various orientation patterns and the method for
calculating orientation score.

control males to show a decrease in heterosexual erotic orientation during non-mating season 3 may have been related to the
fact that all control females were non-pregnant and some could
have passed through weak estrous periods [98]. Finally, concerning social orientation, both groups of males began the
study showing strong isosexuality. While the castrated males
continued to be isosexually oriented, the post-pubertal controls showed a gradual shift toward heterosexuality that was
marked by non-mating season swings back toward same-sex interactions.

<u>Females</u>
The behavioral data for the control and experimental females are summarized in table XVII. As shown in the table, the

control females were more aggressive (both toward one another and toward males) than the experimental females. In addition, the control females showed more isosexual mounts/dyad/hour, more isosexual series-mountings, and more female-male grooming than their experimental counterparts. The experimental females, on the other hand, showed more female-male mounts/dyad/ hour and more female-male series-mountings than the controls (probably due to the actions of female 370). Finally, the experimental females exceeded the control females in both isosexual grooming and isosexual sitting-touching.

The general formula used to calculate female orientation scores was the same as that given for males. Behavioral indices used in the females' calculations were as follows:

1. Agonistic orientation: Female-female fights/dyad/h and heterosexual fights/dyad/h.

2. Erotic orientation: Female-female mounts/dyad/h and male-female mounts/dyad/h.

3. Social orientation (a summation of grooming, sitting-touching, and play): Female-female grooming, sitting-touching, and play/dyad/h and female-male grooming, heterosexual sitting-touching, and heterosexual play/dyad/h.

These indices were similar to those used in the males' calculations with one exception. Females' heterosexual erotic orientation was assumed to be shown by allowing males to mount rather than by mounting males.

As shown in figure 46, the agonistic orientations of both female populations were near equality early in the study, but became increasingly isosexual as the post-operative period progressed. There was little difference between the groups in the strength of agonistic isosexuality, and neither group showed clear seasonal cycles in agonistic orientation.

Within both groups, females' erotic orientation was near equality at the beginning of the study. Early in the post-operative period, the control females began to show seasonal shifts from mating season peaks in heterosexual orientation to non-mating season orientation equality or isosexuality. This pattern disappeared during the last portion of the study, primarily due to the fact that the controls failed to peak in overall male-female mounting during mating seasons 3 and 4 (fig. 21). There were, however, peaks in the controls' male-female series-mountings during these mating periods (fig. 25). This fact suggested that seasonal cyclicity in erotic orientation was continuing.

The erotic orientation of the experimental females fluctuated around equality for the first half of the post-operative period. Thereafter, due to declining frequencies of female-female mounts and relatively steady rates of male-female mounting, their erotic orientation shifted toward heterosexuality.

Finally, the social orientations shown by both female populations tended to be isosexual. The degree of isosexuality was consistently greater among the experimentals than the controls, however, with the intact monkeys showing weak mating season swings toward heterosexuality (fig. 46).

Table XVII. Behavioral summary for females

Variable	Group showing significantly higher value for:		
	Total post-operative period	Summed mating seasons	Summed non-mating seasons
A. Agonism			
1. Female-female fights/d/h*	Controls	Controls	ns**
2. Female-male fights/d/h	Controls	ns	ns
3. Fights/female/h	Controls	Controls	ns
4. Victories/subordinate/h	Controls	Controls	ns
B. Mounting			
1. Female-female mounts/d/h	Controls	Controls	ns
2. Female-male mounts/d/h	Experimentals	Experimentals	ns
3. Mounts/female/h	Controls	Controls	ns
4. Mounts received/female/h	Controls	Controls	ns
C. Series-mounting			
1. Female-female series/d/h	Controls	xxx***	xxx
2. Female-male series/d/h	Experimentals	xxx	xxx

*   Female-female fights/dyad/hour
**   No significant inter-group difference
*** Inter-group difference not tested

Table XVII. (continued)

Variable	Group showing significantly higher value for:		
	Total post-operative period	Summed mating seasons	Summed non-mating seasons
D. Grooming			
1. Female-female episodes/d/h	Experimentals	ns	ns
2. Female-male episodes/d/h	Controls	Controls	Controls
3. Grooming given/female/h	ns	ns	ns
4. Grooming received/fe-male/h	ns	ns	ns
E. Sitting-touching			
1. Female-female episodes/d/h	Experimentals	Experimentals	ns
2. Heterosexual episodes/d/h	ns	ns	ns
3. Episodes/female/h	ns	ns	ns
F. Play			
1. Female-female play/d/h	ns	ns	ns
2. Heterosexual play/d/h	ns	ns	ns
3. Play/female/h	ns	ns	ns

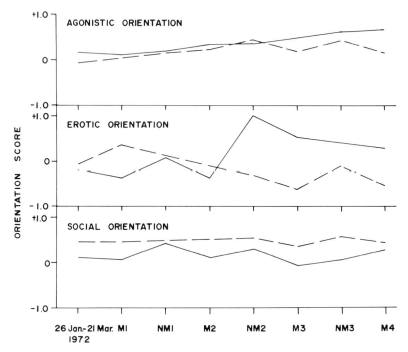

<u>Fig. 46.</u> Orientation patterns for females. Control fe-
males (all) _____, experimental females (all) - - - - - -
- - - - - - - -. Orientation scores range from +1.0 (complete
isosexuality) to -1.0 (complete heterosexuality). See the
text for an explanation of the various orientation patterns
and the method for calculating orientation score.

<u>The Behavior of Rhesus Monkey Juveniles</u>

It is clear that the manipulations of the present study
affected the behavioral rates and orientations of the experimental monkeys, and that males were more strongly affected than
females. Some of the behaviors shown by the adult castrated
males were very similar to those characteristic of much younger monkeys. This fact led us to theorize that these animals
had retained portions of the juvenile behavioral syndrome. In
order to evaluate this idea, and to gain a better understanding of sex differences in behavioral development among rhesus
monkeys, information from other studies of rhesus juveniles
will be examined.

Males

Few longitudinal studies of rhesus social development have been conducted to date. SADE's [147] investigation of the behaviors of seven Cayo Santiago animals (four males and three females) is, perhaps, the best. All seven monkeys were born into one social group during 1961 and were observed intermittently from birth through four years of age.

SADE found that within a free-ranging rhesus group, a young male's early grooming and body contact relationships usually involve close relatives, especially his mother and siblings (see BERMAN [15] for similar data). As the male matures, he expands his interaction sphere to include peers and unrelated adult females. Grooming orientations for SADE's males (calculated from his sociograms) were heterosexual between one to four years of age (range of mean orientation values from -0.24 to -0.11), reflecting relationships with mother, sisters, and unrelated adult females (often mother's close associates). Body contact orientations began as heterosexual but became increasingly isosexual with increasing age (yearlings -0.19, two-year-olds -0.05, four-year-olds 0.11), indicating the development of relationships among male peers. The strongest isosexual orientations shown by SADE's males involved their playful interactions. The males wrestled most often with brothers or male peers, and their mean orientation scores for play ranged from 0.34 as yearlings to 0.57 as two-year-olds, and finally to 0.72 as four-year-olds. When grooming, body contact, and wrestling data were combined to produce a "total social" orientation score (as was done with the data from the present study), SADE's males were found to shift from a slightly heterosexual orientation as yearlings ($\bar{X}$ = -0.03) toward increasing isosexuality as they grew older ($\bar{X}$ = 0.14 for two-year-olds and 0.20 for four-year-olds).

Other studies of rhesus monkeys (generally conducted on corral or laboratory-housed animals) have produced results which agree with SADE's findings. HANBY [70] reported that immature males spent more time in proximity to male peers than females did with peers of either sex. Similarly, RUPPENTHAL and his colleagues [145] reported that within their captive rhesus group, males from 4-36 months of age spent between 51-73% of their "peer interaction time" with other males. With regard to play, several studies have noted that young males play more frequently than young females [16, 100, 127, 145, 160] and agree with SADE [147] that most male play is isosexual [140, 145, 160]. HARLOW and MEARS [73] observed that play becomes increasingly a male-male activity early in the second year of life, primarily because males play more roughly than females. In view of the fact that juvenile males not only play more roughly, but are also better play-fighters than juvenile females (i.e. more likely to achieve the advantageous "on-top" or "behind" positions [160]), the high frequencies of male-male play seen in rhesus groups may reflect both a male preference for same-sex partners and a tendency by females to drop out of vigorous play sessions [7]. Play by males appears to peak in

frequency during the second year of life and then to decline
[100, 127, 140, 160].

As noted by ROSE and his co-workers [139] in their review
of primate agonism, we currently lack good data on the chronological development of aggressive behavior among rhesus monkeys. It is generally agreed, however, that as infants and
juveniles, males are more aggressive than like-aged females
[73, 121]. Further, at least one study has reported a tendency
for the threatening behavior of both infant males and infant
females to be mostly toward same-sex targets [73].

Immature rhesus males mount other monkeys far more frequently than do immature females [16, 127; LOY and LOY, unpub.
obs.]. Furthermore, among captive animals tested in peer
groups, there seems to be an increase in males' mounting rates
during the first two years of life, with plateau levels being
reached at around 24 months [127]. This tendency for males to
mount more frequently than females may be related to higher
levels of male aggressiveness, since most mounts between immature monkeys feature the dominant animal as the mounter [100].
Interestingly, while infant rhesus males generally display an
isosexual play orientation, they have been reported to prefer
to mount female peers rather than each other [61].

Strong evidence for isosexual orientation among infant and
juvenile rhesus males was presented by SUOMI, SACKETT, and
HARLOW [159]. When laboratory-reared males 9-24 months of age
were tested in a self-selection circus situation, they showed
a clear preference for same-sex peers. Isosexual orientation
toward peers began to decline after 24 months, and orientation
became primarily heterosexual between 38-44 months of age.

In summary, as infants and young juveniles, rhesus males
develop strong friendly relations with females (primarily
mother, sisters, and other related females) through grooming
and body contact, and with males (primarily siblings and peers)
through body contact and play. Immature males frequently fight
with and mount peers of both sexes. Nonetheless, most aggression is directed toward other males, while females are the preferred recipients of mounts. As males mature, their grooming
relations with female relatives (especially mother) usually
weaken as the monkeys expand their sphere of relationships to
include unrelated males and females. At the same time, male-
male play relations become stronger as males cluster to engage
in vigorous play-fighting. Most males experience the dissolu-
tion of the social relations of immaturity when they emigrate
from their natal group sometime after reaching sexual maturity
[47, 88]. Within their new groups, the now adult males develop friendly relations primarily with adult females (through
sex and grooming [46, 98]) and direct most of their aggression
toward other adult males [45].

The control males of the present study followed closely
the developmental sequence outlined above, given the limita-
tions imposed by an atypical group composition and the impos-
sibility of emigration. As the intact males matured, they experienced a decline in friendly relations with other males and

developed seasonal cyclicity in social orientation. Additionally, the control males showed the seasonal development of strong heterosexual erotic orientations and peaks in male-female series-mountings, as well as rank-related exclusion of some males from mating activities.

In comparison to the controls, as they grew into adulthood, the castrated males experienced less of a decline in friendly isosexual relations. Male-male play frequencies remained relatively high, as did rates of male-male grooming and sitting-touching. This continuation of strong inter-male friendships into chronological adulthood appeared to be an example of extended or retained juvenility among the castrates. Although they maintained a strongly isosexual social orientation, the adult castrates mounted both males and females frequently. Their strongest erotic bonds were, however, apparently isosexual, a fact reflected by their rates of male-male series-mounting and inter-male penile manipulation. It seems likely that the continuation of close social relations among the adult castrates predisposed them to, or at least facilitated, the development of their unusually strong isosexual erotic orientations.

Females

The three free-ranging females observed by SADE [147] all showed isosexual grooming orientations between one and four years of age (X̄ scores: 0.39 for yearlings, 0.41 for two-year-olds, 0.27 for four-year-olds). The females' most frequent grooming partners tended to be either mother or an older sister. Body contact relations were similar. The females interacted primarily with older female relatives and showed mean sitting-touching orientations of 0.60 as yearlings, 0.23 as two-year-olds, and 0.40 as four-year-olds.

SADE's females engaged in just slightly more heterosexual than isosexual play as yearlings, showing a mean play orientation of -0.06. A year later, their play orientation had become even more heterosexual, as shown by a score of -0.25. In a vivid demonstration of the abrupt cessation of female play at adulthood, none of the three monkeys were observed to play as four-year-olds [147].

The overall social orientations (combining grooming, body contact, and play) of SADE's females were isosexual for all three observation years. As yearlings, their social orientation averaged 0.31, due to near sexual equality in play and a strong grooming/body contact orientation toward female relatives. As two-year-olds, the grooming and contact orientation toward related females continued, but was tempered somewhat by increased heterosexual play, producing a social orientation of 0.13. With the near disappearance of play at sexual maturity, social orientation was once again shaped mainly by isosexual grooming and touching relationships and averaged 0.34 when the subjects were four years old.

Data from several other studies of juvenile rhesus females are available. BERNSTEIN and DRAPER [16] noted that in captivity juvenile females mount less, play less, and sit-touching others more than do juvenile males. RUPPENTHAL and his associates [145] reported that in their captive group infant and juvenile females showed more affiliative behaviors (e.g. groom, huddle) than males, and most affiliative interactions were female-female. Rough and tumble play by females peaks in frequency during the first two years of life and then begins to decline [100, 127, 160], apparently because females drop out as males begin to play roughly [7]. At this time, according to HARLOW and MEARS [73], "... females, left to their own devices, enjoy feminine company more often and begin to reflect the separation of the sexual roles" (p. 197). Interestingly, within free-ranging groups, juvenile females show more isosexual play than expected by chance, even though the majority of their play is with males [160].

As noted earlier, immature rhesus females appear to be less aggressive than like-aged males, and, at least as infants within captive groups of peers, to threaten other females more often than they threaten males [73].

Data on females' preferences in the self-selection circus indicate that when the monkeys are 9-16 months of age, they prefer same-sex stimulus animals. Their preferences shift between 16-25 months, however, and after 25 months of age, females prefer male peers [159].

The pattern of social development typically followed by rhesus females can be summarized as follows. As infants and young juveniles, females develop (through grooming and body contact) strong social relations with mother, sisters, and other female relatives [15]. These isosexual social bonds persist beyond sexual maturity and, since rhesus females do not usually leave their natal group, bind closely the adult animal to her matrilineal genealogy [146]. As females grow, their interaction spheres, of course, do expand to include non-relatives as social partners. Young females play frequently with male and female peers, but as they become older juveniles, females tend to drop out of or avoid the vigorous play of male peers. Juvenile females show considerable interest in infants (this interest is not matched by juvenile males) and attempt to hold and groom them [94]. This interest in infants may promote further development of isosexual social relations since juveniles often groom or sit-touch new mothers in order to be near their infants [94]. In addition to their frequent isosexual social interactions, adult females show seasonal fluctuations in heterosexual social behaviors, engaging in more friendly interactions with adult males during the mating season than during the non-mating (birth) season [46]. The erotic orientation of adult females likewise shows a seasonal pattern. Heterosexual orientation and activity increase markedly during the mating season [98]. Most aggression by adult females is directed toward other members of their age-sex class [45].

Our control females followed closely the normal course of

female development. Since older, related females were not present in their group, they formed strong isosexual social relations with female peers and these bonds persisted into adulthood. Play by the control females declined strongly as they matured, and was rare during adulthood. Social relations with male peers were generally somewhat weaker than inter-female bonds, but showed some intensification during the mating seasons once the monkeys were adults. In addition, the control females' erotic orientation showed seasonal cycles of increasing and decreasing heterosexuality. Finally, most fighting by intact females was isosexual.

Despite their ovariectomies, the experimental females showed most of the features of normal behavioral development. Their primary social orientations were isosexual as youngsters and remained so into adulthood. Play declined strongly as a function of age, and most aggression was directed toward other females. There were, however, some departures from normal development. For example, the ovariectomized females did not develop seasonal cyclicity in either erotic or social orientation. Erotic orientation shifted from pre-operative equality to moderate heterosexuality during adulthood (fig. 46), primarily because male-female mounting remained a relatively frequent occurrence while female-female mounting became rare.

Determinants of the Juvenile-to-Adult Behavioral Transition

The preceding discussion identified several behavioral changes which occur as rhesus monkeys grow from immaturity to adulthood. Some of these developmental patterns were affected by the conditions of the present study, and others were not. The remainder of this chapter will address the related problems of identifying the factors that normally direct behavioral maturation in rhesus macaques, and isolating the variables that caused our experimental monkeys' behavior to deviate from species-typical patterns.

A deceptively attractive theory is that the behavioral changes which characterize the juvenile-to-adult transition are primarily the result of hormonal activation [57, 63]. Whether hormones directly affect particular behaviors, or influence orientation patterns which in turn affect behavior, is not known. The data presented by SUOMI and his colleagues [159] provide strong support for the notion that the hormonal changes at puberty are related to dramatic shifts in orientation toward peers. In their study, socially unsophisticated females and males changed from preferences for same-sex peers to opposite-sex peers at approximately 25 and 40 months of age, respectively. These ages are very near those reported for puberty in females [27] and males [38, 134], and thus the data imply hormonal involvement.

Unfortunately, a close examination of the actual behaviors shown by juvenile and pubertal monkeys provides only moderate

support for the hormonal activation theory. For example, although male-male play is frequent among juveniles and almost
non-existent among adults, the decline in play frequency apparently begins before the hormonal changes of puberty [100, 140,
160]. In addition, under certain circumstances, intact, pre-
pubescent males may precociously display strong heterosexual
erotic orientation and adult sexual behavior, including series-
mounting estrous females [100, 140], while neo-natally castrated males may develop the heterosexual mount pattern complete
with intromission and ejaculation [19]. In sum, while it is
tempting to use the hormonal activation theory to explain many
of the juvenile-to-adult behavior changes, and while such an
explanation may be valid for certain data [159], in most cases
there are reasons to think that other factors may be involved.
These non-hormonal behavioral determinants are usually found
within the animal's social environment [57].

Social factors probably account adequately for many of the
unusual characteristics of the present study's experimental
monkeys. The adult castrated males' failure to develop a
strong heterosexual erotic orientation and the full copulatory
pattern could have been related to the lack of APR females in
their group rather than to their low testosterone levels. It
has been shown that stimuli from estrous females can cause a
recrudescence of heterosexual orientation and copulatory behavior in intact adult rhesus males during the non-mating season [165, 167]. In addition, BIELERT [19] reported that three
of nine neo-natally castrated rhesus males who were periodically allowed to live in groups including two or three intact
female peers showed intromissions and intra-vaginal ejacula-
tions between 36-51 months of age. In light of these data, it
is reasonable to assume that our castrated males might have developed a more heterosexual erotic orientation and copulated
if there had been more than one fully attractive, proceptive,
and receptive female in their group. The special attention
that the castrates gave to female 370 supports this assumption
(see chapter V).

A second feature of our castrated males' behavior which
could have been affected by their social environment was the
retention into adulthood of strong isosexual social relations
(shown by relatively frequent play, grooming, and sitting-
touching). Inter-male social relations among the controls
could have been affected adversely by the passage of all males
through puberty and the beginning of adult mating behavior. As
noted in chapter V, the control males competed for access to
estrous females and success was linked to dominance rank. Although competition was subtle (there were no mating season
peaks in male-male fighting), it may have been sufficient to
cause a reduction in the frequency of friendly interactions
among the control males. The castrated males, on the other
hand, having a paucity of APR females for whom to compete, experienced considerably less weakening of their inter-male
social bonds than did the controls.

The castrated males' failure to develop a strong hetero-

sexual erotic orientation and their concurrent retention of
close friendly relations with other males may have combined to
produce an unusual degree of isosexual eroticism. The experimental males showed significantly higher frequencies of isosexual series-mounting and inter-male penile manipulation than
the controls. The castrates did continue to show an occasional
male-female series-mounting, however, and so their erotic orientation is best described as relatively isosexual, rather than
as exclusively isosexual.

The control females' development of seasonal cycles of
estrus/mating and slightly more heterosexual social orientation
may have been examples of hormonally activated behavior and
orientation shifts. While rhesus males respond to estrous females by becoming sexually active, the opposite is apparently
not true. Quiescent females do not show significant increases
in sexual behavior or ovarian activity when exposed to sexually
active males [168]. Rather, females probably begin their mating season sexual cycles in response to environmental variables
(e.g. decreasing daylength [166, 169]), with inter-female stimulation aiding the process once it starts [168]. Given their
lack of ovaries, the experimental females (with the exception
of 370) could not respond hormonally to environmental stimuli
and thus showed no clear seasonal changes in series-mounting or
social orientation.

Summary

The castrated males were more aggressive, more active
mounters, and more socially active than the control males.
Most of the castrates' interactions (of all sorts) were with
other males. The retention of strong isosexual social relations into adulthood by the castrated males was probably due
to their lack of APR female groupmates and the associated male-
male sexual competition, rather than to their low testosterone
levels. Apparently due (at least partially) to their continuing male-male friendships, the castrates developed a high degree of isosexual eroticism.

The control females fought more and mounted each other
more often than did the experimental females. In addition, the
control females (but not the experimental females) showed mating season shifts toward increased heterosexual social orientation and copulation. The development of these seasonal
cycles may have been the result of hormonal activation of behavior.

XI. Hormones and Society: Some Closing Remarks

In his classic work, ZUCKERMAN [184] concluded that two factors, sexual attraction and dominance, are the main determinants of the behavior and social organization of nonhuman primates. He considered sexual attraction to be the more important variable, calling it the "...main factor that determines social grouping in subhuman primates..." (p. 31). He theorized that, as adults, nonhuman primates experience an uninterrupted reproductive life, with some degree of heterosexual attraction operating at all times, and he linked fluctuations in attraction and interaction to hormonal cycles.

Numerous studies have investigated dominance, sexual attraction, and sexual behavior among primates during the half century since the publication of ZUCKERMAN's book. Space does not permit a thorough review of those studies, although many are cited in earlier chapters. Studies of wild groups and intact groups held in captivity have provided baseline data on the dominance networks, mating systems, and annual hormonal and behavioral cycles of several species. Experimental studies have focused on dominance and sexual behavior, and have been especially helpful in increasing our knowledge of the hormonal influences on behavior. Unfortunately, both types of studies have drawbacks as well as advantages. Naturalistic investigations present the observer with a bewildering assortment of potentially relevant variables--social, physiological, and environmental--only a few of which can be monitored during a single study. Due to the complexity of the system, it is very difficult to isolate individual behavioral determinants for close study. The experimental approach, on the other hand, is generally reductive, manipulating the animals' social and physical environments and their physiology so that one or two variables can be carefully studied. Good examples of the experimental approach can be found among the numerous laboratory studies of hormones and sexual behavior. In these studies, the social setting is often reduced to a pair of animals, and the subjects' physiological states are simplified by gonadectomy followed by the administration of exogenous test substances. While this approach undoubtedly highlights the behavioral effects of hormones, the data thus produced may be difficult to extrapolate to the normal social setting, since hormonal effects can be radically altered by social variables [57, 69, 143].

Our study tried to find a middle ground between the complexity of truly naturalistic studies and the relative simplicity of laboratory investigations. Thus social complexity was simultaneously maintained by working with groups of animals and reduced by limiting the groups to only two age-classes. In a drastic simplification of the subjects' physiological states, all experimental monkeys were gonadectomized. As shown in the previous chapters, this design allowed some insights into the monkeys' behaviors and relationships, and their transition from juvenility to adulthood. There were still numerous instances, however, when we were unable to say whether certain behaviors were influenced primarily by hormones (or the lack of them), or social factors, or some combination of the two.

At least two other studies besides ours have worked with groups of gonadectomized primates. COCHRAN and PERACHIO [33] studied three unisexual trios (one of males and two of females) of adult rhesus monkeys, all of whom had been castrated or ovariectomized. They found evidence for hormonal (exogenous dihydrotestosterone propionate) effects on aggression, dominance rank, and male sexual behavior. Perhaps significantly, the one trio which did not experience dominance changes in response to hormone treatment had been established for a longer period than the other groups, suggesting that social factors overrode hormonal effects in this case.

Undoubtedly the most difficult gonadectomy study to interpret is ALVAREZ' [8, 9] investigation of the effects of castration and ovariectomy on squirrel monkeys (Saimiri sciureus). ALVAREZ formed two social groups of adults who had no history of prior contact with each other. Each group contained three males and three females, and after a short period of baseline observations, all members of one group (the experimentals) were gonadectomized. Observations were then made for a period with the experimental monkeys agonadal and receiving no exogenous hormones. The final phase of the study involved the administration of estradiol benzoate and testosterone propionate to the experimental females and males, respectively. The control monkeys remained intact throughout the study. They did not undergo sham surgery at the time of the experimentals' operations, but they were given placebo injections of sesame oil to match the experimentals' hormone injections.

ALVAREZ found that despite the fact that his groups had been formed during the breeding season (as evidenced by the males' "fatted" condition), both experimentals and controls showed a sex-segregated pattern of social organization during the pre-operative, baseline period. Sex-segregation was reflected by the average distances between females and between males being much smaller than the average male-female distance, and by isosexual body contact being much more frequent than heterosexual touching. This basic pattern of sex-segregation has since been shown to exist year-round among squirrel monkeys, although there is an increase in sex-integration during the breeding season [10, 11, 12, 34, 35, 48, 91, 105, 108].

ALVAREZ' experimental monkeys responded to gonadectomy by becoming more heterosexually oriented. Male-female social distance decreased sharply and heterosexual body contact increased. At the same time, female-female social distance in-

creased, and both inter-male and inter-female body contacts
dropped in frequency. The males seemed to have been somewhat
more affected by gonadectomy and to have contributed slightly
more to the increase in sex-integration than the females.

The post-operative shift from strongly sex-segregated toward increased sex-integration suggested to ALVAREZ that gonadectomy had somehow produced an increase in heterosexual interest or attraction. While presumably the changes were due to
the loss of the gonadal hormones (implying that the presence of
those hormones was responsible for the sex-segregation pattern
[10]), this possibility was weakened by the failure of hormone
therapy to restore the sex-segregated pattern [8]. More recently, several studies have demonstrated that male attraction
toward females is positively correlated with both estrogen and
testosterone levels [10, 35, 108], thus casting further doubt
on the theory that low hormone levels produce sex-integration
in Saimiri.

As demonstrated in the earlier chapters, the low hormone
levels of our experimental rhesus monkeys were correlated with
a strongly sex-segregated pattern (i.e. isosexual orientation).
Although to a great extent this pattern was probably due to
social factors (e.g. the lack of both APR females and sexual
competition among the castrated males), many of the social
factors in turn were largely determined by the monkeys' lack of
gonadal hormones. Among our control monkeys, the females apparently showed hormonally activated seasonal shifts toward
heterosexual erotic and social orientations. Increased male
heterosexuality may have then been triggered by the presence of
APR females. In general, the results of the present study support ZUCKERMAN's conclusion that the gonadal hormones promote
heterosexual attraction and interaction.

In conclusion, our study was undertaken to gain new information concerning the influence exerted by the gonadal hormones on rhesus monkey behavior and social organization. Although we found considerable evidence that hormones act as behavioral potentiators [57], in most instances the final form
taken by behavior was strongly affected by social factors as
well. The interactions of physiological and social factors in
the production of orientation and behavior are probably more
complex during the juvenile-to-adult transition than at any
other time in a monkey's life. This study focused primarily on
that transitional period (along with the early adult years) and
we hope it has provided some new insights into the processes
and mechanisms of rhesus behavioral maturation. As with all
scientific studies, if this work stimulates discussion and further research, it will have served a worthwhile purpose.

XII. References

1 AKERS, J.S. and CONAWAY, C.H.: Female homosexual behavior
       in Macaca mulatta. Arch. sex. Behav. 8: 63-80 (1979).
2 ALDIS, O.: Play fighting; pp. 99-106 (Academic Press,
       New York 1975).
3 ALEXANDER, B.K. and ROTH, E.M.: The effects of acute
       crowding on aggressive behavior of Japanese monkeys.
       Behaviour 39: 73-90 (1971).
4 ALLEN, M.L. and LEMMON, W.B.: Orgasm in female primates.
       Am. J. Primatol. 1: 15-34 (1981).
5 ALTMANN, J.: Observational study of behavior: sampling
       methods. Behaviour 49: 227-263 (1974).
6 ALTMANN, S.A.: A field study of the sociobiology of rhesus
       monkeys, Macaca mulatta. Ann. N.Y. Acad. Sci. 102:
       338-435 (1962).
7 ALTMANN, S.A.: Sociobiology of rhesus monkeys. III. The
       basic communication network. Behaviour 32: 17-32
       (1968).
8 ALVAREZ, F.: Effects of sex hormones on the social organi-
       zation of the squirrel monkey, Saimiri sciureus; un-
       published doctoral dissertation (Tulane University,
       New Orleans 1969).
9 ALVAREZ, F.: Comportamiento social y hormonas sexuales en
       "Saimiri sciureus"; Monografias de la Estacion
       Biologica de Doñana; no. 2 (Consejo Superior de
       Investigaciones Cientificas, Madrid 1973).
10 ANDERSON, C.O. and MASON, W.A.: Hormones and social be-
       havior of squirrel monkeys (Saimiri sciureus). I.
       Effects of endocrine status of females on behavior
       within heterosexual pairs. Horm. Behav. 8: 100-106
       (1977).
11 BALDWIN, J.D.: The social behavior of adult male squirrel
       monkeys (Saimiri sciureus) in a seminatural environ-
       ment. Folia primatol. 9: 281-314 (1968).
12 BALDWIN, J.D.: The social organization of a semifree-
       ranging troop of squirrel monkeys (Saimiri sciureus).
       Folia primatol. 14: 23-50 (1971).
13 BEACH, F.A.: Factors involved in the control of mounting
       behavior by female mammals; in DIAMOND, Reproduction
       and sexual behavior; pp. 83-131 (Indiana University
       Press, Bloomington 1968).
14 BEACH, F.A.: Sexual attractivity, proceptivity, and recep-
       tivity in female mammals. Horm. Behav. 7: 105-138
       (1976).

15 BERMAN, C.M.: The ontogeny of social relationships with group companions among free-ranging rhesus monkeys. I. Social networks and differentiation. Anim. Behav. 30: 149-162 (1982).

16 BERNSTEIN, I.S. and DRAPER, W.A.: The behaviour of juvenile rhesus monkeys in groups. Anim. Behav. 12: 84-91 (1964).

17 BERNSTEIN, I.S.; GORDON, T.P., and PETERSON, M.: Role behavior of an agonadal alpha-male rhesus monkey in a heterosexual group. Folia primatol. 32: 263-267 (1979).

18 BERTRAND, M.: The behavioral repertoire of the stumptail macaque Biblthca primatol. 11: 197-203 (S. Karger, Basel 1969).

19 BIELERT, C.F.: The effects of early castration and testosterone propionate treatment on the development and display of behavior patterns by male rhesus monkeys; unpublished doctoral dissertation (Michigan State University, East Lansing 1974).

20 BIELERT, C.; CZAJA, J.S.; EISELE, S.; SCHEFFLER, G.; ROBINSON, J.A., and GOY, R.W.: Mating in the rhesus monkey (Macaca mulatta) after conception and its relationship to oestradiol and progesterone levels throughout pregnancy. J. Reprod. Fertil. 46: 179-187 (1976).

21 BIRCH, H.G. and CLARK, G.C.: Hormonal modifications of social behavior. II. The effects of sex-hormone administration on the social dominance status of the female-castrate chimpanzee. Psychosom. Med. 8: 320-331 (1946).

22 BLIZZARD, R.M.; THOMPSON, R.G.; BAGHDASSARIAN, A.; KOWARSKI, A.; MIGEON, C.J., and RODRIGUEZ, A.: The interrelationship of steroids, growth hormone, and other hormones on pubertal growth; in GRUMBACH, GRAVE and MAYER, Control of the onset of puberty; pp. 342-366 (John Wiley and Sons, New York 1974).

23 BREUGGEMAN, J.A.: The function of adult play in free-ranging Macaca mulatta; in SMITH, Social play in primates; pp. 169-191 (Academic Press, New York 1978).

24 CARPENTER, C.R.: Sexual behavior of free ranging rhesus monkeys (Macaca mulatta): specimens, procedures, and behavioral characteristics of estrus, J. comp. Psychol. 33: 113-142 (1942).

25 CARPENTER, C.R.: Sexual behavior of free ranging rhesus monkeys (Macaca mulatta): periodicity of estrus, homosexual, autoerotic and nonconformist behavior. J. comp. Psychol. 33: 143-162 (1942).

26 CARPENTER, C.R.: Tentative generalizations on the grouping behavior of nonhuman primates. Human Biology 26: 269-276 (1954).

27 CATCHPOLE, H.R. and VAN WAGENEN, G.: Reproduction in the rhesus monkey, Macaca mulatta; in BOURNE, The rhesus

   monkey; vol. 2, pp. 117-140 (Academic Press, New
   York 1975).

28 CHANNING, C.P.; FOWLER, S.; ENGEL, B., and VITEK, K.:
   Failure of daily injections of ketamine HCl to adversely alter menstrual cycle length, blood estrogen,
   and progesterone levels in the rhesus monkey. Proc.
   Soc. exp. Biol. Med. 155: 615-619 (1977).

29 CHEVALIER-SKOLNIKOFF, S.: Male-female, female-female, and
   male-male sexual behavior in the stumptail monkey,
   with special attention to the female orgasm. Arch.
   sex. Behav. 3: 95-116 (1974).

30 CHEVALIER-SKOLNIKOFF, S.: Homosexual behavior in a laboratory group of stumptail monkeys (Macaca arctoides):
   forms, contexts, and possible social functions. Arch.
   sex. Behav. 5: 511-527 (1976).

31 CHEVALIER-SKOLNIKOFF, S. and POIRIER, F.E.: Primate bio-
   social development (Garland Publishing Co., New York
   1977).

32 CLARK, G. and BIRCH, H.G.: Hormonal modifications of
   social behavior: I. The effect of sex-hormone administration on the social behavior of a male-castrate
   chimpanzee. Psychosom. Med. 7: 321-329 (1945).

33 COCHRAN, C.A. and PERACHIO, A.A.: Dihydrotestosterone
   propionate effects on dominance and sexual behaviors
   in gonadectomized male and female rhesus monkeys.
   Horm. Behav. 8: 175-187 (1977).

34 COE, C.L. and ROSENBLUM, L.A.: Sexual segregation and its
   ontogeny in squirrel monkey social structure. J. hum.
   Evol. 3: 551-561 (1974).

35 COE, C.L. and ROSENBLUM, L.A.: Annual reproductive strat-
   egy of the squirrel monkey (Saimiri sciureus). Folia
   primatol. 29: 19-42 (1978).

36 COE, C.L.; MENDOZA, S.P.; DAVIDSON, J.M.; SMITH, E.R.;
   DALLMAN, M.F., and LEVINE, S.: Hormonal response to
   stress in the squirrel monkey (Saimiri sciureus).
   Neuroendocrinology 26: 367-377 (1978).

37 CONAWAY, C.H. and KOFORD, C.B.: Estrous cycles and mating
   behavior in a free-ranging band of rhesus monkeys.
   J. Mammal. 45: 577-588 (1965).

38 CONAWAY, C.H. and SADE, D.S.: The seasonal spermatogenic
   cycle in free ranging rhesus monkeys. Folia primatol.
   3: 1-12 (1965).

39 COUNT, E.W.: Being and becoming human: essays on the
   biogram (D. Van Nostrand Co., New York 1973).

40 CZAJA, J.A. and BIELERT, C.: Female rhesus sexual behavior
   and distance to a male partner: relation to stage of
   the menstrual cycle. Arch. sex. Behav. 4: 583-597
   (1975).

41 CZAJA, J.A.; EISELE, S.G., and GOY, R.W.: Cyclical changes
   in the sexual skin of female rhesus: relationships to
   mating behavior and successful artificial insemina-
   tion. Fed. Proc. 34: 1680-1684 (1975).

42 DICZFALUSY, E. and STANDLEY, C.C.: The use of non-human

primates in research on human reproduction (World Health Organization, Stockholm 1972).

43 DIXSON, A.F.: Androgens and aggressive behavior in primates: a review. Aggressive Behav. 6: 37-67 (1980).

44 DOLHINOW, P.J. and BISHOP, N.: The development of motor skills and social relationships among primates through play; in HILL, Minnesota symposia on child psychology; vol. 4, pp. 141-198 (University of Minnesota Press, Minneapolis 1970).

45 DRICKAMER, L.C.: Quantitative observation of behavior in free-ranging Macaca mulatta: methodology and aggression. Behaviour 55: 209-236 (1975).

46 DRICKAMER, L.C.: Quantitative observations of grooming behavior in free-ranging Macaca mulatta. Primates 17: 323-335 (1976).

47 DRICKAMER, L.C. and VESSEY, S.H.: Group changing in free-ranging male rhesus monkeys. Primates 14: 359-368 (1973).

48 DUMOND, F.V.: The squirrel monkey in a seminatural environment; in ROSENBLUM and COOPER, The squirrel monkey; pp. 87-145 (Academic Press, New York 1968).

49 DUVALL, S.W.; BERNSTEIN, I.S., and GORDON, T.P.: Paternity and status in a rhesus monkey group. J. Reprod. Fertil. 47: 25-31 (1976).

50 EATON, G.G.: Male dominance and aggression in Japanese macaque reproduction; in MONTAGNA and SADLER, Reproductive behavior; Adv. behav. Biol. 11: 287-297 (Plenum, New York 1974).

51 Eaton, G.G.: Longitudinal studies of sexual behavior in the Oregon troop of Japanese macaques; in MCGILL, DEWSBURY and SACHS, Sex and behavior; pp. 35-59 (Plenum, New York 1978).

52 EATON, G.G. and RESKO, J.A.: Plasma testosterone and male dominance in a Japanese macaque (Macaca fuscata) troop compared with repeated measures of testosterone in laboratory males. Horm. Behav. 5: 251-259 (1974).

53 EHRHARDT, A.A.: Androgens in prenatal development: behavior changes in nonhuman primates and men. Adv. Biosci. 13: 153-162 (1974).

54 ERWIN, J. and MAPLE, T.: Ambisexual behavior with male-male anal penetration in male rhesus monkeys. Arch. sex. Behav. 5: 9-14 (1976).

55 FAIMAN, C. and WINTER, J.S.D.: Gonadotropins and sex hormone patterns in puberty: clinical data; in GRUMBACH, GRAVE and MAYER, Control of the onset of puberty; pp. 32-61 (John Wiley and Sons, New York 1974).

56 GADPAILLE, W.J.: Cross-species and cross-cultural contributions to understanding homosexual activity. Arch. Gen. Psychiatry 37: 349-356 (1980).

57 GOLDFOOT, D.A.: Sociosexual behaviors of nonhuman primates during development and maturity: social and hormonal relationships; in SCHRIER, Behavioral primatology:

advances in research and theory; pp. 139-184
(Lawrence Ehrlbaum, Hillsdale, N.J. 1977).

58 GORDON, T.P.; BERNSTEIN, I.S., and ROSE, R.M.: Social and
seasonal influences on testosterone secretion in the
male rhesus monkey. Physiol. Behav. $\underline{21}$: 623-627
(1978).

59 GORDON, T.P.; ROSE, R.M., and BERNSTEIN, I.S.: Seasonal
rhythm in plasma testosterone levels in the rhesus
monkey (Macaca mulatta): a three year study. Horm.
Behav. $\underline{7}$: 229-243 (1976).

60 GORDON, T.P.; ROSE, R.M.; GRADY, C.L., and BERNSTEIN, I.S.:
Effects of increased testosterone secretion on the
behavior of adult male rhesus living in a social
group. Folia primatol. $\underline{32}$: 149-160 (1979).

61 GOY, R.W. and GOLDFOOT, D.A.: Experiential and hormonal
factors influencing development of sexual behavior
in the male rhesus monkey; in SCHMITT and WORDEN,
The neurosciences, third study program; pp. 571-581
(MIT Press, Cambridge 1974).

62 GOY, R.W. and GOLDFOOT, D.A.: Neuroendocrinology: animal
models and problems of human sexuality. Arch. sex.
Behav. $\underline{4}$: 405-420 (1975).

63 GOY, R.W. and MCEWEN, B.S.: Sexual differentiation of the
brain (MIT Press, Cambridge 1980).

64 GOY, R.W. and PHOENIX, C.H.: The effects of testosterone
propionate administered before birth on the development of behavior in genetic female rhesus monkeys;
in SAWYER and GORSKI, Steroid hormones and brain
function; pp. 193-205 (University of California
Press, Berkeley 1972).

65 GOY, R.W.; WALLEN, K., and GOLDFOOT, D.A.: Social factors
affecting the development of mounting behavior in
male rhesus monkeys; in MONTAGNA and SADLER, Reproductive behavior; Adv. behav. Biol. $\underline{11}$: 223-247
(Plenum, New York 1974).

66 HAFEZ, E.S.E.: Comparative reproduction of nonhuman primates (Charles C. Thomas, Springfield 1971).

67 HALL, K.R.L. and DEVORE, I.: Baboon social behavior; in
DEVORE, Primate behavior: field studies of monkeys
and apes; pp. 53-110 (Holt, Rinehart and Winston,
New York 1965).

68 HANBY, J.P.: Male-male mounting in Japanese monkeys
(Macaca fuscata). Anim. Behav. $\underline{22}$: 836-849 (1974).

69 HANBY, J.P.: Social factors affecting primate reproduction; in MONEY and MUSAPH, Handbook of sexology;
pp. 461-484 (Elsevier/North-Holland Biomedical Press,
Amsterdam 1977).

70 HANBY, J.P.: Relationships in six groups of rhesus monkeys. II. Dyads. Am. J. phys. Anthrop. $\underline{52}$: 565-575
(1980).

71 HANBY, J.P.; ROBERTSON, L.T., and PHOENIX, C.H.: The
sexual behavior of a confined troop of Japanese
macaques. Folia primatol. $\underline{16}$: 123-143 (1971).

72 HARLOW, H.F. and HARLOW, M.K.: The affectional systems; in SCHRIER, HARLOW and STOLLNITZ, Behavior of nonhuman primates; vol. 2, pp. 287-334 (Academic Press, New York 1965).

73 HARLOW, H.F. and MEARS, C.: The human model: primate perspectives; pp. 189-213 (V.H. Winston and Sons, Washington, D.C. 1979).

74 HARTMAN, C.G.: Studies in the reproduction of the monkey Macacus (Pithecus) rhesus, with special reference to menstruation and pregnancy; in Contr. Embryol. 23: 1-161 (Carnegie Inst., Washington 1932).

75 HAUSFATER, G.: Dominance and reproduction in baboons (Papio cynocephalus): a quantitative analysis; Contr. Primatol. vol. 7 (Karger, Basel 1975).

76 HERBERT, J. and TRIMBLE, M.R.: Effect of oestradiol and testosterone on the sexual receptivity and attractiveness of the female rhesus monkey. Nature 216: 165-166 (1967).

77 HESS, D.L. and RESKO, J.A.: The effects of progesterone on the patterns of testosterone and estradiol concentrations in the systemic plasma of the female rhesus monkey during the intermenstrual period. Endocr. 92: 446-453 (1973).

78 HINDE, R.A. and SPENCER-BOOTH, Y.: The behaviour of socially living rhesus monkeys in their first two and a half years. Anim. Behav. 15: 169-196 (1967).

79 HOLLOWAY, R.L.: Primate aggression, territoriality, and xenophobia (Academic Press, New York 1974).

80 HUTCHINS, M. and BARASH, D.P.: Grooming in primates: implications for its utilitarian function. Primates 17: 145-150 (1976).

81 HURME, V.O. and VAN WAGENEN, G.: Basic data on the emergence of permanent teeth in the rhesus monkey (Macaca mulatta). Proc. Am. phil. Soc. 105: 105-140 (1961).

82 JOHNSON, D.F. and PHOENIX, C.H.: Hormonal control of female sexual attractiveness, proceptivity, and receptivity in rhesus monkeys. J. comp. physiol. Psychol. 90: 473-483 (1976).

83 JOSLYN, W.D.: Androgen-induced social dominance in infant female rhesus monkeys. J. Child Psychol. Psychiat. 14: 137-145 (1973).

84 KAPLAN, E.L. and MEIER, P.: Nonparametric estimation from incomplete observations. Am. Stat. Assoc. J. 53: 457-481 (1958).

85 KAUFMANN, J.H.: A three-year study of mating behavior in a freeranging band of rhesus monkeys. Ecology 46: 500-512 (1965).

86 KAUFMANN, J.H.: Social relations of adult males in a free-ranging band of rhesus monkeys; in ALTMANN, Social communication among primates; pp. 73-98 (University of Chicago Press, Chicago 1967).

87 KEVERNE, E.B.: Sexual receptivity and attractiveness in

the female rhesus monkey; in ROSENBLATT, HINDE, SHAW and BEER, Advances in the study of behavior; pp. 155-200 (Academic Press, New York 1976).

88 KOFORD, C.B.: Population changes in rhesus monkeys: Cayo Santiago, 1960-1964. Tulane Studies in Zoology 13: 1-7 (1966).

89 LANCASTER, J.B.: Sex and gender in evolutionary perspective; in KATCHADOURIAN, Human sexuality: a comparative and developmental perspective; pp. 51-80 (University of California Press, Berkeley 1979).

90 LANCASTER, J.B. and LEE, R.B.: The annual reproductive cycle in monkeys and apes; in DEVORE, Primate behavior: field studies of monkeys and apes; pp. 486-513 (Holt, Rinehart and Winston, New York 1965).

91 LEGER, D.W.; MASON, W.A., and MUNKENBECK FRAGASZY, D.: Sexual segregation, cliques, and social power in squirrel monkey (Saimiri) groups. Behaviour 76: 163-181 (1981).

92 LERESCHE, L.A.: Dyadic play in hamadryas baboons. Behaviour 57: 190-205 (1976).

93 LINDBURG, D.G.: A field study of the reproductive behavior of the rhesus monkey (Macaca mulatta); unpublished doctoral dissertation (University of California, Berkeley 1967).

94 LINDBURG, D.G.: The rhesus monkey in North India: an ecological and behavioral study; in ROSENBLUM, Primate behavior: developments in field and laboratory research; vol. 2, pp. 1-106 (Academic Press, New York 1971).

95 LINDBURG, D.G.: Grooming behavior as a regulator of social interactions in rhesus monkeys; in CARPENTER, Behavioral regulators of behavior in primates; pp. 124-148 (Bucknell University Press, Lewisburg 1973).

96 LOIZOS, C.: Play behaviour in higher primates: a review; in MORRIS, Primate ethology; pp. 176-218 (Aldine, Chicago 1967).

97 LOY, J.D.: Estrus behavior of free-ranging rhesus monkeys (Macaca mulatta): a study of continuity and variability; unpublished doctoral dissertation (Northwestern University, Evanston 1969).

98 LOY, J.: Estrous behavior of free-ranging rhesus monkeys. Primates 12: 1-31 (1971).

99 LOY, J.: The descent of dominance in Macaca: insights into the structure of human societies; in TUTTLE, Socioecology and psychology of primates; pp. 153-180 (Mouton, The Hague/Paris 1975).

100 LOY, J. and LOY, K.: Behavior of an all-juvenile group of rhesus monkeys. Am. J. phys. Anthrop. 40: 83-96 (1974).

101 LOY, J.; LOY, K.; PATTERSON, D.; KEIFER, G., and CONAWAY, C.: The behavior of gonadectomized rhesus monkeys. I. Play; in SMITH, Social play in primates; pp. 49-78 (Academic Press, New York 1978).

102 LUTTGE, W.G.: The role of gonadal hormones in the sexual
        behavior of the rhesus monkey and human: a liter-
        ature review. Arch. sex. Behav. 1: 61-88 (1971).
103 MARSDEN, H.M.: Agonistic behaviour of young rhesus monkeys
        after changes induced in social rank of their moth-
        ers. Anim. Behav. 16: 38-44 (1968).
104 MASON, J.W.; KENION, C.C.; COLLINS, D.R.; MOUGEY, E.H.;
        JONES, J.A.; DRIVER, G.C.; BRADY, J.V., and BEER, B.:
        Urinary testosterone response to 72-h avoidance ses-
        sions in the monkey. Psychosom. Med. 30: 721-732
        (1968).
105 MASON, W.A.: Differential grouping patterns in two species
        of South American monkey; in WHITE, Ethology and
        psychiatry; pp. 153-169 (University of Toronto Press,
        Toronto 1974).
106 MAZUR, A.: A cross-species comparison of status in small
        established groups. Am. Sociological Review 38: 513-
        530 (1973).
107 MAZUR, A.: Effects of testosterone on status in primate
        groups. Folia primatol. 26: 214-226 (1976).
108 MENDOZA, S.P.; LOWE, E.L.; RESKO, J.A., and LEVINE, S.:
        Seasonal variations in gonadal hormones and social
        behavior in squirrel monkeys. Physiol. Behav. 20:
        515-522 (1978).
109 MICHAEL, R.P.: Hormonal factors and aggressive behaviour
        in the rhesus monkey; in Influence of hormones on
        the nervous system; Proc. Int. Soc. Psychoneuro-
        endocrinology, Brooklyn 1970; pp. 412-423 (Karger,
        Basel 1971).
110 MICHAEL, R.P.; HERBERT, J., and WELEGALLA, J.: Ovarian
        hormones and grooming behaviour in the rhesus monkey
        (Macaca mulatta) under laboratory conditions. J.
        Endocr. 36: 263-279 (1966).
111 MICHAEL, R.P.; HERBERT, J., and WELEGALLA, J.: Ovarian
        hormones and the sexual behavior of the male rhesus
        monkey (Macaca mulatta) under laboratory conditions.
        J. Endocr. 39: 81-98 (1967).
112 MICHAEL, R.P. and SAAYMAN, G.: Sexual performance and the
        timing of ejaculation in male rhesus monkeys (Macaca
        mulatta). J. comp. physiol. Psychol. 64: 213-218
        (1967).
113 MICHAEL, R.P. and WILSON, M.: Changes in the sexual be-
        haviour of male rhesus monkeys (M. mulatta) at
        puberty. Folia primatol. 19: 384-403 (1973).
114 MICHAEL, R.P. and WILSON, M.: Effects of castration and
        hormone replacement in fully adult male rhesus mon-
        keys (Macaca mulatta). Endocr. 95: 150-159 (1974).
115 MICHAEL, R.P.; WILSON, M., and PLANT, T.M.: Sexual behav-
        iour of male primates and the role of testosterone;
        in MICHAEL and CROOK, Comparative ecology and behav-
        iour of primates; pp. 235-313 (Academic Press, London
        1973).
116 MICHAEL, R.P.; WILSON, M.I., and ZUMPE, D.: The bisexual

behavior of female rhesus monkeys; in FRIEDMAN, RICHART and VANDE WIELE, Sex differences in behavior; pp. 399-412 (John Wiley and Sons, New York 1974).

117 MICHAEL, R.P. and ZUMPE, D.: Sexual initiating behaviour by female rhesus monkeys (<u>Macaca</u> <u>mulatta</u>) under laboratory conditions. Behaviour <u>36</u>: 168-186 (1970).

118 MICHAEL, R.P. and ZUMPE, D.: Annual cycles of aggression and plasma testosterone in captive male rhesus monkeys. Psychoneuroendocrinology. <u>3</u>: 217-220 (1978).

119 MICHAEL, R.P. and ZUMPE, D.: Relation between the seasonal changes in aggression, plasma testosterone and the photoperiod in male rhesus monkeys. Psychoneuroendocrinology. <u>6</u>: 145-158 (1981).

120 MIRSKY, A.: The influence of sex hormones on social behavior in monkeys. J. comp. physiol. Psychol. <u>48</u>: 327-335 (1955).

121 MITCHELL, G.: Behavioral sex differences in nonhuman primates (Van Nostrand Reinhold Co., New York 1979).

122 MITCHELL, G. and TOKUNAGA, D.H.: Sex differences in nonhuman primate grooming. Behav. Processes <u>1</u>: 335-345 (1976).

123 NISWENDER, G.D. and SPIES, H.G.: Serum levels of luteinizing hormone, follicle-stimulating hormone and progesterone throughout the menstrual cycle of rhesus monkeys. J. clin. Endocr. Metab. <u>37</u>: 326-328 (1973).

124 OVERPECK, J.G.; COLSON, S.H.; HOHMANN, J.R.; APPLESTINE, M.S., and REILLY, J.F.: Concentrations of circulating steroids in normal prepubertal and adult male and female humans, chimpanzees, rhesus monkeys, rats, mice and hamsters: a literature survey. J. Toxicol. envir. Health <u>4</u>: 785-803 (1978).

125 PACKER, C.: Male dominance and reproductive activity in <u>Papio</u> <u>anubis</u>. Anim. Behav. <u>27</u>: 37-45 (1979).

126 PHOENIX, C.H.: Sexual behavior in rhesus monkeys after vasectomy. Science <u>179</u>: 493-494 (1973).

127 PHOENIX, C.H.: Prenatal testosterone in the nonhuman primate and its consequences for behavior; in FRIEDMAN, RICHART and VANDE WIELE, Sex differences in behavior; pp. 19-32 (John Wiley and Sons, New York 1974).

128 PHOENIX, C.H.; GOY, R.W., and RESKO, J.A.: Psychosexual differentiation as a function of androgenic stimulation; in DIAMOND, Reproduction and sexual behavior; pp. 33-49 (Indiana University Press, Bloomington 1968).

129 PHOENIX, C.H. and JENSEN, J.N.: Ejaculation by male rhesus in the absence of female partners. Horm. Behav. <u>4</u>: 231-238 (1973).

130 PHOENIX, C.H.; SLOB, A.K., and GOY, R.W.: Effects of castration and replacement therapy on sexual behavior of adult male rhesuses. J. comp. physiol. Psychol. <u>84</u>: 472-481 (1973).

131 PLANT, T.M.; ZUMPE, D.; SAULS, M., and MICHAEL, R.P.: An annual rhythm in the plasma testosterone of adult

male rhesus monkeys maintained in the laboratory. J. Endocr. 62: 403-404 (1974).

132 PURI, C.P.; PURI, V., and ANAND KUMAR, T.C.: Serum levels of testosterone, cortisol, prolactin and bioactive luteinizing hormone in adult male rhesus monkeys following cage-restraint or anaesthetizing with ketamine hydrochloride. Acta Endocrinologica 97: 118-124 (1981).

133 RALEIGH, M.J.; FLANNERY, J.W., and ERVIN, F.R.: Sex differences in behavior among juvenile vervet monkeys (Cercopithecus aethiops sabaeus). Behav. & Neural Biol. 26: 455-465 (1979).

134 RESKO, J.A.: Plasma androgen levels of the rhesus monkey: effects of age and season. Endocr. 81: 1203-1212 (1967).

135 RESKO, J.A.: Sex steroids in adrenal effluent plasma of the ovariectomized rhesus monkey. J. clin. Endocr. Metab. 33: 940-948 (1971).

136 RESJO, J.A.: The relationship between fetal hormones and the differentiation of the central nervous system in primates; in MONTAGNA and SADLER, Reproductive behavior; Adv. behav. Biol. 11: 211-222 (Plenum, New York 1974).

137 RESKO, J.A. and PHOENIX, C.H.: Sexual behavior and testosterone concentrations in the plasma of the rhesus monkey before and after castration. Endocr. 91: 499-503 (1972).

138 ROBINSON, J.A.; SCHEFFLER, G. EISELE, S.G., and GOY, R.W.: Effects of age and season on sexual behavior and plasma testosterone and dihydrotestosterone concentrations of laboratory-housed male rhesus monkeys (Macaca mulatta). Biol. Reprod. 13: 203-210 (1975).

139 ROSE, R.M.; BERNSTEIN, I.S.; GORDON, T.P., and CATLIN, S.F.: Androgens and aggression: a review and recent findings in primates; in HOLLOWAY, Primate aggression, territoriality, and xenophobia; pp. 275-304 (Academic Press, New York 1974).

140 ROSE, R.M.; BERNSTEIN, I.S.; GORDON, T.P., and LINDSLEY, J.G.: Changes in testosterone and behavior during adolescence in the male rhesus monkey. Psychosom. Med. 40: 60-70 (1978).

141 ROSE, R.M.; GORDON, T.P., and BERNSTEIN, I.S.: Diurnal variation in plasma testosterone and cortisol in rhesus monkeys living in social groups. J. Endocr. 76: 67-74 (1978).

142 ROSE, R.M.; HOLADAY, J.W., and BERNSTEIN, I.S.: Plasma testosterone, dominance rank and aggressive behaviour in male rhesus monkeys. Nature 231: 366-368 (1971).

143 ROWELL, T.E.: Female reproduction cycles and social behavior in primates; in LEHRMAN, HINDE and SHAW, Advances in the study of behavior; vol. 4, pp. 69-105 (Academic Press, New York 1972).

144   ROWELL, T.: The social behaviour of monkeys (Penguin, Harmondsworth, Middlesex 1972).

145   RUPPENTHAL, G.C.; HARLOW, M.K.; EISELE, C.D.; HARLOW, H.F., and SUOMI, S.J.: Development of peer interactions of monkeys reared in a nuclear-family environment. Child Develop. 45: 670-682 (1974).

146   SADE, D.S.: Some aspects of parent-offspring and sibling relations in a group of rhesus monkeys, with a discussion of grooming. Am. J. phys. Anthrop. n.s. 23: 1-18 (1965).

147   SADE, D.S.: Ontogeny of social relations in a group of free-ranging rhesus monkeys (Macaca mulatta Zimmerman); unpublished doctoral dissertation (University of California, Berkeley 1966).

148   SADE, D.S.: Determinants of dominance in a group of free-ranging rhesus monkeys; in ALTMANN, Social communication among primates; pp. 99-114 (University of Chicago Press, Chicago 1967).

149   SADE, D.S.: Sociometrics of Macaca mulatta. I. Linkages and cliques in grooming matrices. Folia primatol. 18: 196-223 (1972).

150   SADE, D.S.; CUSHING, K.; CUSHING, P.; DUNAIF, J.; FIGUEROA, A.; KAPLAN, J.R.; LAUER, C.; RHODES, D., and SCHNEIDER, J.: Population dynamics in relation to social structure on Cayo Santiago. Yearbook of Physical Anthrop. (1976) 20: 253-262 (1977).

151   SEYFARTH, R.M.: A model of social grooming among adult female monkeys. J. theor. Biol. 65: 671-698 (1977).

152   SIMONDS, P.E.: The bonnet macaque in South India; in DEVORE, Primate behavior: field studies of monkeys and apes; pp. 175-196 (Holt, Rinehart and Winston, New York 1965).

153   SMITH, D.G.: The association between rank and reproductive success of male rhesus monkeys. Am. J. Primatol. 1: 83-90 (1981).

154   SMITH, E.O.: A historical view of the study of play: statement of the problem; in SMITH, Social play in primates; pp. 1-32 (Academic Press, New York 1978).

155   SOUTHWICK, C.H.: Aggressive behaviour of rhesus monkeys in natural and captive groups; in GARATTINI and SIGG, Aggressive behaviour; Proc. Symp. on Biol. of Aggress. Behav., Milan, May 1968; pp. 32-43 (Excerpta Medica, Amsterdam 1969).

156   SOUTHWICK, C.H. and SIDDIQI, M.R.: The role of social tradition in the maintenance of dominance in a wild rhesus group. Primates 8: 341-353 (1967).

157   SPARKS, J.: Allogrooming in primates: a review; in MORRIS, Primate ethology; pp. 148-175 (Aldine, Chicago 1967).

158   STRUHSAKER, T.T.: Behavior of vervet monkeys (Cercopithecus aethiops); University of California Publications in Zoology; vol. 82 (University of California Press, Berkeley 1967).

159   SUOMI, S.J.; SACKETT, G.P., and HARLOW, H.F.: Development

of sex preference in rhesus monkeys. Develop. Psych.
$\underline{3}$: 326-336 (1970).

160 SYMONS, D.: Play and aggression: a study of rhesus monkeys (Columbia University Press, New York 1978).

161 SYMONS, D.: The evolution of human sexuality (Oxford
University Press, Oxford 1979).

162 TERRY, R.L.: Primate grooming as a tension reduction
mechanism. J. Psychol. $\underline{76}$: 129-136 (1970).

163 TRIMBLE, M.R. and HERBERT, J.: The effect of testosterone
or oestradiol upon the sexual and associated behaviour of the adult female rhesus monkey. J. Endocr.
$\underline{42}$: 171-185 (1968).

164 VANDENBERGH, J.G.: Hormonal basis of sex skin in male
rhesus monkeys. Gen. compar. Endocr. $\underline{5}$: 31-34 (1965).

165 VANDENBERGH, J.G.: Endocrine coordination in monkeys: male
sexual responses to the female. Physiol. Behav. $\underline{4}$:
261-264 (1969).

166 VANDENBERGH, J.G.: Environmental influences on breeding
in rhesus monkeys; in Symp. IVth int. congr. primat.;
vol. 2: primate reproductive behavior; pp. 1-19
(Karger, Basel 1973).

167 VANDENBERGH, J.G. and DRICKAMER, L.C.: Reproductive coordination among free-ranging rhesus monkeys. Physiol.
Behav. $\underline{13}$: 373-376 (1974).

168 VANDENBERGH, J.G. and POST, W.: Endocrine coordination in
rhesus monkeys: female responses to the male.
Physiol. Behav. $\underline{17}$: 979-984 (1976).

169 VAN HORN, R.N.: Seasonal reproductive patterns in primates.
Prog. reprod. Biol. $\underline{5}$: 181-221 (Karger, Basel 1980).

170 VAN WAGENEN, G.: Maturity induced by testosterone in the
young male monkey. Fed. Proc. $\underline{6}$: 219 (1947).

171 VAN WAGENEN, G.: Accelerated growth with sexual precocity
in female monkeys receiving testosterone propionate.
Endocr. $\underline{45}$: 544-546 (1949).

172 VAN WAGENEN, G. and CATCHPOLE, H.R.: Physical growth of
the rhesus monkey (Macaca mulatta). Am. J. phys.
Anthrop. n.s. $\underline{14}$: 245-273 (1956).

173 VAN WAGENEN, G. and HURME, V.O.: Effect of testosterone
propionate on permanent canine tooth eruption in the
monkey (Macaca mulatta). Proc. Soc. exp. Biol. Med.
$\underline{73}$: 296-297 (1950).

174 VARLEY, M. and SYMMES, D.: The hierarchy of dominance in
a group of macaques. Behaviour $\underline{27}$: 54-75 (1966).

175 WALLEN, K.; BIELERT, C., and SLIMP, J.: Foot clasp mounting in the prepubertal rhesus monkey: social and
hormonal influences; in CHEVALIER-SKOLNIKOFF and
POIRIER, Primate bio-social development; pp. 439-461
(Garland, New York 1977).

176 WALLEN, K.; GOLDFOOT, D.A., and GOY, R.W.: Peer and maternal influences on the expression of foot-clasp
mounting by juvenile male rhesus monkeys. Dev.
Psychobiol. $\underline{14}$: 299-309 (1981).

177 WICKINGS, E.J. and NIESCHLAG, E.: Testosterone production

and metabolism in laboratory-maintained male rhesus monkeys. Int. J. Fertil. <u>22</u>: 56-59 (1977).

178 WICKLER, W.: The sexual code (Anchor Press/Doubleday, Garden City 1973).

179 WILKS, J.W.; HODGEN, G.D., and ROSS, G.T.: Endocrine characteristics of ovulatory and anovulatory menstrual cycles in the rhesus monkey; in HAFEZ, Human ovulation; pp. 205-218 (Elsevier/North-Holland Biomedical Press, Amsterdam 1979).

180 WILSON, A.P. and BOELKINS, C.: Evidence for seasonal variation in aggressive behaviour by <u>Macaca mulatta</u>. Anim. Behav. <u>18</u>: 719-724 (1970).

181 WILSON, A.P. and VESSEY, S.H.: Behavior of free-ranging castrated rhesus monkeys. Folia primatol. <u>9</u>: 1-14 (1968).

182 WILSON, M.E.; GORDON, T.P., and CHIKAZAWA, D.: Female mating relationships in rhesus monkeys. Am. J. Primatol. <u>2</u>: 21-27 (1982).

183 WOLFHEIM, J.H.: Sex differences in behavior in a group of captive juvenile talapoin monkeys (<u>Miopithecus talapoin</u>). Behaviour <u>63</u>: 110-128 (1977).

184 ZUCKERMAN, S.: The social life of monkeys and apes (Harcourt, Brace and Co., New York 1932).

185 ZUCKERMAN, S.; VAN WAGENEN, S.G., and GARDINER, R.H.: The sexual skin of the rhesus monkey. Proc. zool. Soc. Lond. <u>108</u>: 385-401 (1938).

186 ZUMPE, D. and MICHAEL, R.P.: Ovarian hormones and female sexual invitations in captive rhesus monkeys (<u>Macaca mulatta</u>). Anim. Behav. <u>18</u>: 293-301 (1970).

187 ZUMPE, D. and MICHAEL, R.P.: Redirected aggression and gonadal hormones in captive rhesus monkeys (<u>Macaca mulatta</u>). Anim. Behav. <u>18</u>: 11-19 (1970).

XIII. Subject Index